AWAKEN

THE PHYSICS OF ASCENSION

DAVID ASH

WITH MATTHEW NEWSOME &
SUSAN SAILLARD-THOMPSON

Published by:

Kima Global Publishers,
50, Clovelly Road,
Clovelly, 7975
South Africa

ISBN 978-1-928234-23-4
eISBN 978-1-928234-24-1

Publisher's web site: www.kimabooks.com
Author's web site: www.davidash.co

Author's Patreon page: www.patreon.com/David_Ash

Cover art by Argonaut Design Ltd.

Other books by David Ash:

The Tower of Truth (with Anna Mary Ash) CAMSPRESS, 1978
The Vortex: Key to Future Science, (with Peter Hewitt) Gateway, 1990
The New Science of the Spirit, The College of Psychic Studies, 1995
Activation for Ascension, Kima Global Publishers, 1995
The Power of Puja, Puja Power Publications, 2004
The Power of Physics, Puja Power Publications, 2005
The New Physics of Consciousness, Kima Global Publishers, 2007
The Role of Evil in Human Evolution, Kima Global Publishers, 2007
The Vortex Theory, Kima Global Publishers, 2015
Continuous Living, Kima Global Publishers, 2015

DEDICATION

To Amma and my family and friends.

I'd rather be one of the devotees of God than one of the straight, so-called sane or normal people who just don't understand that man is a spiritual being, that he has a soul.

George Harrison

When religion, shorn of its superstitions, traditions and unintelligent dogmas, shows its conformity with science, then there will be a great unifying cleansing force in the world which will sweep before it all wars, disagreements, discords and struggles – and then will mankind be united in the power of the Love of God.

'Abdu'l-bahá

Spirituality without quantum physics is an incomplete picture of reality.

Dalai Lama

NON-SCIENTISTS ARE RECOMMENDED TO READ THE BOOKS IN REVERSE ORDER.

TABLE OF CONTENTS

ACKNOWLEDGEMENTS

Lulu, Josephine, and Nick, thank you for lending me *Sapiens: A Brief History of Humankind* which inspired me to write this book. Suzie Saillard-Thompson, thank you for your love, your listening and your many contributions to *Awaken*. I could not have written it without you. Vincent, I thank you for the Doris Stokes autobiography. I am grateful as always, to my father and mother, Sir George Trevelyan, Nigel Blair and Peter Hewitt. I thank Nigel Calder for the Richard Feynman quotes and I wish to express my appreciation to the Hindu Temple in Neasden, London for providing information about the origin of science in India and also for the quotes from luminaries about India. Jez and Jason of Argonaut Design thank you for your generosity in designing the cover of *Awaken* and Agi for the awesome lotus. Priscilla Husband, Anne McEwen, Jimena, Zophia, Ashen, Richard and Suzie, thank you for your kindness and hospitality while I was writing *Awaken*. Linda and Sacha thank you for your help too. Matthew Newsome I thank you for your editing and your many additions and for your love and support and that of your family and also Nalini, thank you for your companionship, light and inspiration over many years and for bringing Suzie, Matthew, Priscilla and Jimena into my life. Robin and Nadine Beck thank you for publishing my books for over twenty five years. Rebecca and Anna thank you for your quotes and Jessica thank you for your drawing of Democritus. Jana, thank you for your drawing of the Phoenix rising. Anna, I thank you for providing your charming drawings. Thank you, Sam and Fenton for your inputs. I thank Anna's sister Jo and my sisters Mary and Jenny and Bruce and my brothers Steven, Simon, Peter and Richard and also Kim, Hillary and Liesbeth. I thank Suzie's family especially Phillippe for his input. I thank all those in my very large family especially Ondine, Lily, Lorna and Raphael who I have not already mentioned and my wide circle of friends who support, love and believe in me now and have done so for many years.

INTRODUCTION

I was inspired to write AWAKEN when I read *Sapiens: A brief History of Humankind*[1] by Yuval Noah Harari. In his more recent book *21 Lessons for the 21st Century*,[2] Harari contends that everything we believe in, apart from science, is a story. In the book he draws on practically every sphere of human belief, especially religion, to reveal the extent to which we have been taken in by stories. But he failed to include the story he believes in, the story of materialism underlying science. This is common of most sapiens; we are quick to question the beliefs of others but we are not so quick to question our own beliefs.

In *21 Lessons* Harari revealed his belief in materialism when he wrote: "In itself the Universe is only a meaningless hodgepodge of atoms". In that statement Harari summarised the atomic hypothesis which is the core principle in the philosophy of materialism that was promulgated by Democritus two and a half thousand years ago and is still believed by most mainstream scientists and philosophers today.

In *Awaken* I denounce the common belief in materialism and present a new story for science based on the idea that subatomic particles are not corpuscles of material substance but are *whirlpools of light*. This simple idea enabled me to explain how, in nuclear physics, light and matter can so easily interchange. Over a period of fifty years I developed this alternative to materialism into a full blown scientific theory.

In my research I discovered that the quark theory is a farrago of nonsense and there is critical evidence in physics that has been covered up for the best part of ninety years. Like an unexploded bomb, this inconvenient truth threatens to blow quantum physics apart. I can tell you now there is more reason to believe

1 Harari Y.N. *Sapiens: A Brief History of Humankind* Vintage 2014

2 Harari Y.N. *21 Lessons for the 21st. Century* Jonathan Cape 2018.

in angels and fairies than in quarks and the virtual particles of quantum mechanics.

From my understanding that particles of matter are whirlpools of light I realised that we could exist in one of a number of parallel realities based on layers of energy all coinciding with each other. However, because of a set of self evident rules these other realities may be invisible and intangible to us. Some of them might even be a part of us; in fact they might be who we really are. As a scientist, I was confronted by the possibility that soul and spirit might not be fables; they might be based on fact.

It has become fashionable today to dismiss the supernatural in the name of science but the science we think is right might be wrong. Future generations may discount a lot of the science we believe in today as story much as our generation is dismissing religions as stories.

Scientists say that science is evidence based when in fact it is selective evidence based. Most scientists cherry pick what is to be researched and what is to be dismissed in terms of their faith in materialism rather than in the merits of the evidence before them. As I tell in AWAKEN there is a mountain of evidence for the paranormal which is ignored in mainstream science, because of the general belief in the philosophy of materialism that nothing exists but what we can see and touch. Materialists are biased in their belief that there is no meaning or purpose in the Universe. They believe everything is a consequence of totally random interactions of minute particles which they think exist of themselves. Science is prejudiced by this belief no less than religions are blinkered by religious beliefs.

Even though we may find fault in religions and they are undoubtedly layered with stories, we are in danger of losing incredibly rich traditions and clues to many truths that may underlie them. In my research I have discovered that many of the religious teachings that liberal, secular people are inclined to dismiss as *hocus pocus* could be based on fact in terms of my new understanding of energy and my appreciation of why Albert Einstein put so much emphasis on the speed of light. My work on the whirlpools of light has shed light on many things in religion and spirituality as well as in science.

I warn against an over-reliance on materialistic beliefs and values because materialism is destroying the Earth. Materialism is also closing our minds to the possibility of higher dimensions of reality in the Universe. Indeed it may be blocking us from realising our multidimensional potential. Materialism could be leading us toward extinction. We are deluded by materialism. Materialism is an illusion, it is a myth. This is proven by quantum physics because according to quantum theory everything is made of particles of energy and energy has no mass or material substance; energy is no-thing.

How can energy that is 'no-thing' form solid things like mountains and houses? When I was a teenager I realised something truly amazing while reading Yogic philosophy. The Yogis in ancient India had probed the atom with extraordinary powers of mind over matter. They saw that subatomic particles are not material things but are minute vortices of spinning light. Thousands of years before Einstein the Yogis cracked the greatest enigma in modern physics, how energy forms mass. They realised that materialism is a delusion set up by subatomic spin. Yogis called this Maya: the illusion of forms.

The quantum world seems to be more like a dream than a material reality. Particles of energy appear to be more like thoughts than things. Many quantum physicists believe the Universe is a mind and consider consciousness to be the bedrock of reality. This allows for amazing possibilities.

Einstein established that the world we live in is based on energy which is fundamentally movement at the speed of light. In the vortex theory I have predicted there may be worlds of super-energy existing beyond the speed of light, worlds beyond our current range of perception.

If we can accept that our bodies are fundamentally composed of light, in the form of vortices of energy, then the new vortex physics points to the promise of immortality. If we can let go of the illusion of materialism then death may not be the end for us.

If the energy in all the whirlpools of light in every atom of our body were to be accelerated to speeds beyond that of light then our bodies would disappear out of this world and appear alive, intact and somewhat surprised in a world of super energy. This is what I predict to be the beginning of a resonance process commonly called *Ascension* which enables bodies to be shifted out of physical space-time into super-physical dimensions. The technology underlying this super-energy resonance process has been alluded to in Star Trek as: "Beam me up Scotty." Sci-fi often precedes Sci-fact.

The worlds of super-energy could be populated by sentient beings. This would explain reports throughout history of super-physical lifeforms such as angels. Some humans may have already gone through the ascension process. This fits with near death experiences where people claim to have found themselves in heaven where they met human beings who have ascended. There are numerous reports of enlightened masters disappearing and re-appearing throughout human history. These immortals have become known as the Ascended Masters. It would make sense that those who have ascended before us would return to our dimension and encourage us to let go of our attachment to materialism so that we can join the 'party in the sky!'

If you find what I am saying unbelievable then I invite you to be a sceptic like me. Question everything you believe and review with an open mind the physics and the super-physics I present herein. Find out for yourself. Our time here is brief. Not all who wander are lost.

AWAKEN is a journey of discovery in seven books. Read them in order as they appeal to you. The first book reveals the many myths in modern theoretical physics. The second book presents the new vortex physics that arises from the Yogic insight into the atom. The cosmology I outline in book three gives a revolutionary vortex account for gravity.

Book four is concerned with consciousness and brings intelligence to the theory of evolution showing how super-energy worlds existing beyond the speed of light could have decisively influenced the origin of life on earth, and how our world can be visited by super-energy beings without our being aware of them.

Book five deals with the issues of soul and spirit from the perspective of the super-energy and includes subjects such as life after death, re-incarnation and near death experiences. Book five also explains how we may be proto-angels on Earth training for eternal life.

Book six explains miracles and magic in terms of super-energy resonance technology and defines the process underlying physical ascension.

Book seven reviews the source of the ascension predictions as they came to me. It also describes the possibility of a planetary transition in terms of a cycle of predictable geo-physical change as endorsed by Albert Einstein. This book concludes with a corroboration of these predictions as foretold by the remarkable Bulgarian master, Peter Deunov, whom Einstein in his own words "bowed down before". This extraordinary prophecy, delivered to the world, in 1944, two days prior to Peter Deunov's death, upholds the core message of AWAKEN.

If they come the sweeping earth changes could be preceded by an exceedingly rare opportunity for humanity to ascend en masse into super-physicality. The call of ascension is to the heart more than the head. At the moment of ascension the light in our hearts will be drawn to the light emanating from a door of light that will appear before us. We will enter it in response to our hearts not our reasoning minds. Now is the time to prepare for this opportunity by living more from our hearts than our minds in the practice of love, gratitude and forgiveness. The reading of AWAKEN may well be the chance for you and many others to awaken to this unparalleled opportunity prior to the forth-coming planetary transition.

BOOK I

MYTHS IN MODERN PHYSICS

"In terms of basic concepts, science has lurched forward on the backs of a few creative individuals who came into prominence when the time was right for them. In many instances their work was actually rejected, at the time of its conception because of the pressure of prevailing opinion and thought, and only became acceptable to a later generation looking for a new direction. All the same, the fundamental concepts of science are rarely questioned, once established. It is very difficult for any brave souls to fly in the face of peer pressure. It is more than their jobs are worth...And those like myself, who come from outside the world of academia are largely ignored by the institutions of science. Those who are not ignored get ostracized and attacked like an invading beetle in a bee-hive. Their presence is emotionally uncomfortable to the majority who tow the party line."[1]

John Davidson

1 Davidson J. *The Secret of the Creative Vacuum,* Daniel 1989

CHAPTER 1
CHALLENGING THE STANDARD THEORY

Today, there is universal confidence in science but science has become a new religion to replace the old. In the new religion the focus of belief and worship is on materialism instead of God.

Scientists and sceptics are caught up in a net of delusion by materialism. They seem to have blind faith in materialism when it has been shown by Einstein's theories and his equation $E=mc^2$ to be a myth. The same is true of other theories in science. Many are mythologies perpetuated more as scientific fundamentalism rather than scientific fact.

As it was with the crushing dogmatism of the medieval church so now science has become infected with an arrogance of mind that goes with presumed certainty of truth. Worst of all, science has succumbed to ignorance. Ignorance is a derivative of the verb to ignore which has, mistakenly, been taken to mean: to lack information when its obvious meaning is: to ignore information.

Science has been used to support ignorance – in the true sense of the word – by pseudo-sceptics who operate, ostensibly, in the name of science. A true sceptic questions everything. Pseudo-sceptics question everything except the philosophy of materialism they believe in. Pseudo-skeptics, acting as 'defenders of the faith' resist with the fervour of fundamentalists anything that doesn't conform to the materialistic world view of science and Western philosophy. If they can't explain away the contradictory evidence to scientific materialism they simply ignore it.

In *Unraveling the Mind of God*,[2] Robert Matthew recorded a comment, typical of 'ignorant' (meaning those who ignore) scientists. It was made by Robert Oppenheimer, the father of the atom bomb, about the work of David Bohm. Bohm was a physicist and author of *The Implicate Order,* a quantum concept that what we take for reality are explicate forms that have unfolded out of an underlying implicate order which is the ground of all being. Oppenheimer said of Bohm's theory, "We can't find anything wrong with it so we will just have to ignore it".

The original purpose of science was to destroy ignorance. Now science, like the medieval Roman Catholic Church, has become pervaded with ignorance.

Confidence in the medieval church collapsed when its cosmology was disputed during the Copernican revolution. When Galileo Galilei observed the moons around Jupiter, through his rudimentary telescope, he disproved Church dogma that the Earth is at the centre of the Universe. That was how Galileo shattered the credibility of Catholic doctrine.

Science is about to suffer a similar crisis in confidence which could lead to a new Copernican revolution. This may happen because physicists failed to take heed of Albert Einstein's warning against placing too much store by the principle of uncertainty proposed by Werner Heisenberg.

It is not Heisenberg's principle, in principle, that is a problem. The uncertainty principle, in itself, is elegant and profound. The concept of uncertainty in all theories, beliefs and philosophies is vital to maintain, in order to ensure the constant flow of ideas otherwise with certainty in belief, progress stagnates, ingenuity is stifled and freedom of thought is lost as people become enslaved by the beliefs.

The problem with Heisenberg's principle was the way physics has been manipulated to protect it from being

2 Matthews, R. *Unraveling the Mind of God* Virgin 1992

disproved. Albert Einstein disliked the uncertainty principle on the grounds that it was impossible to prove wrong.

Einstein fell from grace for questioning the quantum mechanics headed by Heisenberg. However, in 1932, in the decade after Einstein was toppled from the pinnacle of physics, a particle called the *neutron* was discovered at the Cavendish Laboratory at Cambridge, which vindicated him in more ways than one. It led to the atomic bomb, which provided dramatic proof of his equation $E=mc^2$. It also offered an opportunity to test Heisenberg's principle and the outcome was equally explosive. Experimental evidence about the neutron has disproved the uncertainty principle. (This is detailed in Chapter 4.)

Physicists refused to accept the evidence. They buried their heads in the sand when they realised theories in quantum mechanics dependent on the uncertainty principle may have been built on quicksand. To begin with they ignored the evidence that could blow their quantum world apart until quark theory came along. Quark theory was convenient as it enabled physicists to cobble an account together to try to explain the annoying little neutron, so it would no longer pose a threat to the uncertainty principle. However, another fundamental particle, the *proton,* which had also been discovered in Britain, undermined credibility in the quark theory.

If the quantum theory and quark theory are fundamentally flawed, the way may be open to reconsider the paradigm busting implications of the vortex theory. The vortex theory is worth considering as it is able to provide a satisfactory account for both of these fundamental particles in question and overcome the difficulties they present in quantum theory and quark theory.

There is no place for a new theory if the old one is satisfactory. Progress in science is made when fault is found in an established theory that the upstart theory can correct. Then the new theory can replace the old. Such is the revolution of science, as my father, a medical doctor and research scientist summed up when he said to me, "Before you give them a nut to crack, crack their nut."

Quantum mechanics is currently the most successful theory in the history of science and is a main strut in the Standard theory of physics. However advocates of quantum mechanics have been lying to themselves, lying to each other and lying to the world.

Before I detail the damning evidence that derails quantum mechanics I have to lay bare the inadequacies in quark theory. This is because quark theory has been used to dismiss evidence pertaining to the neutron which disproves the uncertainty principle.

CHAPTER 2
QUARK NONSENSE

Before World War II, the contents of the Universe seemed simple. Every atom was formed of just three kinds of particle; the electron, the proton and the neutron.

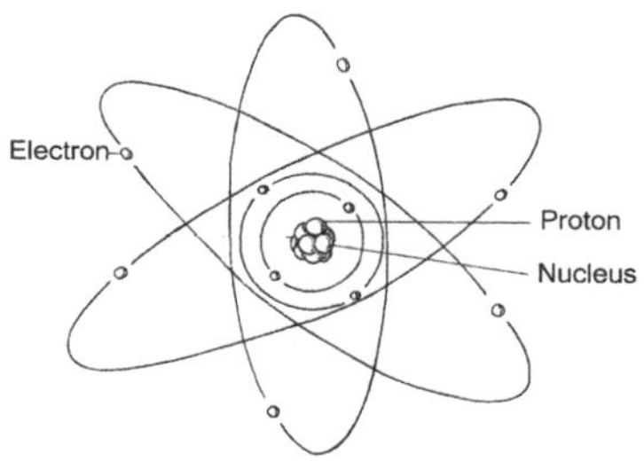

During World War II, the neutron was used to drive the chain reaction in the atomic bombs dropped on Hiroshima and Nagasaki. I call it the *Shiva particle* after Shiva, the Hindu god of destruction.

By the 1960's physicists were confronted with dozens of very short lived heavy particles, which came to light in cosmic rays experiments, or in experiments performed in the new particle accelerators. The newly discovered particles were given Greek names, and collectively they came to be known as the *particle zoo.* Most of the new particles seemed to be more massive relatives of the proton because many of them decayed into a proton. Physicists began to imagine the proton itself was not a truly basic particle. They began to postulate that it was made up of even more fundamental particles.

Just as the atom had been discovered to be made of subatomic particles, it was imagined that subatomic particles were made of yet more elementary particles. These imagined particles came to be called *quarks*.

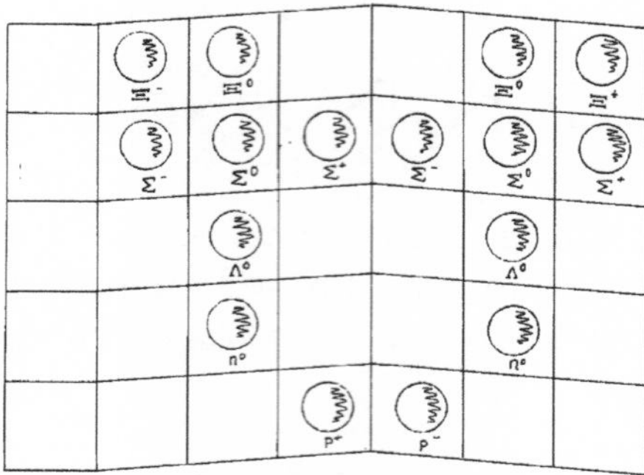

The Particle Zoo

Murray Gell-Mann, of The California Institute of Technology (then in his early thirties), was the author of *Quark Theory*. He declared that protons, neutrons and the newly discovered heavy particles were made out of three quarks. He accounted for the different subatomic particles in terms of combinations of different types of quark.

However, Gell-Mann was unable to explain the difference in life span of the newly discovered particles compared to the proton – discovered in 1917, when Ernest Rutherford split the atom at Manchester University.

The life of a proton has been estimated at a billion, trillion, trillion (10^{33}) years, whereas one ten billionth (10^{-10}) of a second is considered to be a long life for any of the new particles found in high-energy research. As spontaneous proton decays have never been observed, the proton could be treated as being infinitely more stable than any of the heavy particles that have come to light in the particle accelerators, such as the European Centre for Nuclear Research, CERN.

Physicists call the heavy particles *baryons* from the Greek word for heavy. They account for the difference in life span between protons and other baryons in a law called the law of *Conservation of Baryon Numbers*. However, this law doesn't tell us why protons live for so much longer than the other particles; it simply states that they do. It is just one of the many instances in science where a law is proclaimed to describe how something behaves without giving a reason for why it behaves the way it does.

With over 31 million seconds in a year, the difference between the life span of a proton and one of the new particles is a figure in excess of 3×10^{50} (3 with fifty zeros behind it). Mathematicians use the number 10^{50} as a cut off point for probability. If the probability of something happening is less than 1 in 10^{50} then scientists are forced to accept that it never happened. With a difference in stability in excess of 10^{50}, it can be argued that it is improbable that protons and the new short lived baryons are of the same fundamental nature.

Imagine walking down a road between two building sites. On the site to the left, the houses disintegrate as soon as they are built. Within a billionth of a second they are gone. On the site to the right, however, the houses are advertised for sale with a billion, trillion, trillion year guarantee. No one in their right mind would suggest that identical bricks and mortar and construction techniques are employed on both building sites. Surely the first question any sensible person would ask about those houses would be, "Why does one lot last so long when the other lot disappears so quickly?"

The baryon particles are like houses in this analogy. Quarks are equivalent to bricks and the *gluon bonds* which bind them together, represent mortar. Quark theory cannot be applied to protons as well as all the newly discovered particles unless its infinite stability, compared to the incredible instability of the other baryons, is accounted for.

Quark theory could be used to explain the stable protons, or it could be used to account for the unstable new particles, but not both. To use quark theory to explain only protons would be pointless, as the theory was invented to explain the new particles discovered in high energy physics. But if quark theory provided an explanation for all the baryons, apart from the proton, it would be meaningless because the great bulk of the mass of the physical Universe consists of protons, whereas the other baryons – apart from the neutron – have been observed only as anomalies in high-energy experiments.

Physicists have never isolated a quark. All they ever see in high energy experiments are the tracks of debris particles like electron jets or mesons, which they had predicted as break-down products of quarks. They say the existence of quarks has been confirmed by evidence from bombarding protons with high speed electrons. In these experiments, conducted at the Stanford Linear Accelerator (SLAC) in the 1960's, electrons ricocheted as though they hit something hard inside the protons. Physicists concluded the hard things were quarks. These experiments have been repeated at CERN but they did not prove the existence of quarks. The experiments merely showed there was something hard in the proton, but it may not have been a quark, it may have been something else. In the *New Vortex Theory* I explain what that something else could be.

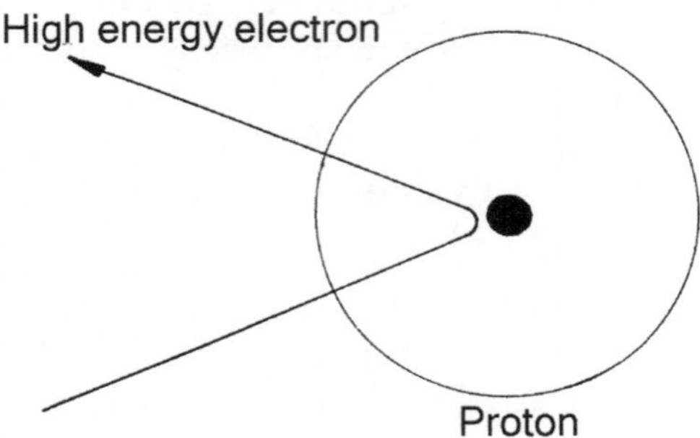

Despite the fact that fractional electric charges have never been observed in nature the existence of fractional charge is assumed in the quark theory. This speculation is in defiance of experimental physics. Also quark theory has deduced that quarks are heavier than the protons they are supposed to form, which led Richard Feynman to say, "The problem of particle masses has been swept in the corner."[1]

1 Calder N., *Key to the Universe: A Report on the New Physics,* BBC Publications 1977

While everything in the world is supposed to be made of quarks, not a single quark has ever been observed in a free state - even though neither expense nor effort has been spared in the attempts to find one. Quark theory would appear to be nonsense, as its name suggests, quark is slang for nonsense in the German language. It could be that quarks do not exist.

The underlying problem with quark theory is the world view of materialism that quark theory represents. Quarks are believed to be the ultimate particles of material substance. The search for quarks is, in reality, a search for the material atoms of Democritus when in reality no such atoms exist.

CHAPTER 3
PROBLEMS WITH THE NEUTRON

In a simple world everything would be made of just two charged particles, the proton and the electron. Electrons are light particles with a negative charge. Protons are much more massive than electrons and they have a positive charge. The third particle in the atom is the neutron. That could be, quite simply, an electron bound to a proton; they are, after all, attracted to each other by their opposite charge attraction.

A neutron is neutral in charge and can be formed out of an electron and a proton (in K-capture). But, unlike the electron and proton, the neutron is unstable. Outside the atomic nucleus, or in a radioactive nucleus, the neutron can fall apart into an electron and proton (in Beta decay). The neutron also has the sum mass of an electron and a proton.

From the experimental evidence one would imagine that the neutron must be an electron bound to a proton. That would explain its neutral charge, because opposite charges cancel each other out.

But the world of physics is not that simple. Physicists refuse to accept that a neutron is an electron bound to a proton. They believe that when an electron and a proton come together, to form a neutron, the electron and proton lose their identity altogether and regain it again when the unstable neutron falls apart. They teach students that the proton is made up of *two up-quarks* and *one down-quark*, the neutron is made up of *two down-quarks* and one up-quark, and the quarks are bound together with *gluon bonds*.[1]

1 Calder N, *Key to the Universe: A Report on the New Physics*, BBC Publications 1977

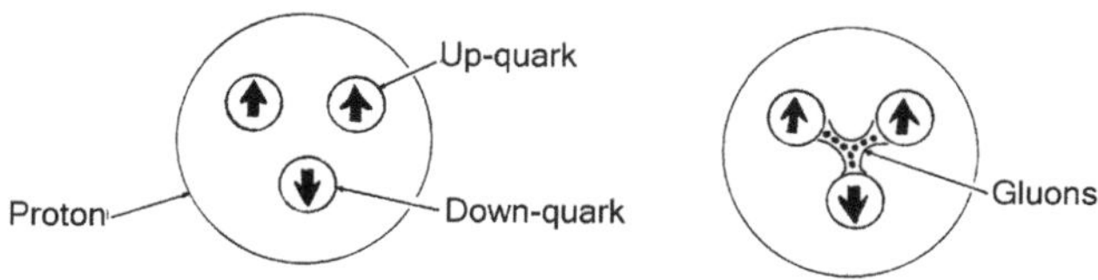

Up-quarks are supposed to have 2/3 charge and down-quarks are supposed to have 1/3 charge. In the proton 2/3 + 2/3 - 1/3 = 1 which, in quark theory, confers upon the proton a unitary charge. In the neutron, however, 1/3 + 1/3 - 2/3 = 0. That gives the neutron zero charge.[2]

According to quark theory, when a proton interacts with an electron, a weak nuclear force comes into play which transforms an

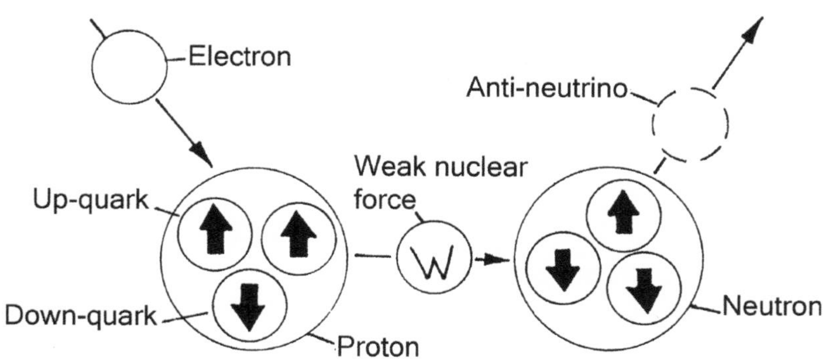

up-quark into a down-quark and the electron into an anti-neutrino

2 Calder N, *Key to the Universe: A Report on the New Physics*, BBC Publications 1977

which is temporarily lost. In this process the electron and proton cease to exist and their place is taken by the neutron and a neutrino. When a neutron decays, the proton reappears with the anti-neutrino which then allows the electron to return and the neutrino to vanish and take away some energy with it as it goes.[3]

The quark theory for neutrons is depicted in a story about a *Mad Professor of Puddings*:

A student in a university department of nutrition baked plums in a pudding then ran round shouting, "Eureka, I have just invented plum pudding!"

His professor was not impressed. He snorted over his spectacles, "That is not a plum pudding you baked. That is a Black forest gateau," he exclaimed.

"But," argued the downcast student, "I mixed plums into my pudding before I baked it and when I weighed it, the weight was that of the plums and the pud and then, when I shook the pudding, out came a plum. It has to be a plum pudding!"

The professor became very angry. "You stupid student, have you learnt nothing of what I have taught you? When you bake plums in a pudding the ingredients change their identity. Due to the action of a weak cooking force, the plums change into cherries and pudding becomes a Black forest gateau."

"Well how do you explain the plum that fell out of the pudding?" retorted the student; "There are no plums in a Black forest gateau!"

"The cherry gateau is unstable," huffed the professor, "after a few minutes it falls apart into its original ingredients, plums and pudding."

How could the student argue? He was speaking to none other than the President of the Royal Society of Puddings. If he wanted to make it in the world and get a good career in the cake

3 Calder N, *Key to the Universe: A Report on the New Physics*, BBC Publications 1977

and pudding industry, he had to accept that a pudding full of plums was a Black forest gateau.

In the story, the plum represents an electron, the pudding mass is the proton, and the plum pudding is the neutron. But the story doesn't end there. Back in the nutrition department all was not going well for the *Mad Professor of Puddings*:

Another student baked plums in a pudding then licked it. He tasted a plum and so established that the plums did not lose their identity when baked in a pudding. The mad professor was very embarrassed by the arrival of that awkward fact. He ranted that the lick was inconclusive and raved that because of the margin of error, in the lick experiment, there was no certainty that anything was actually tasted in the lick. "You didn't taste anything when you licked the pudding", he shouted at the student, "The taste could be exactly zero, in agreement with my theory."

The lick experiment in that story parodied an experiment performed in 1957.[4] In the experiment a weak electric dipole was measured on a neutron, which seemed to indicate neutrons are not entirely neutral, as expected in the standard theory. On one spot the neutron displayed a minute negative charge – around a billion, trillion times weaker than that of a single electron. This appeared to suggest that the neutron could be a bound state of opposite charges, which mostly cancel each other out – as in the atom, which is electrically neutral because it contains equal numbers of oppositely charged particles.

Physicists were quick to dismiss the experiment that revealed the electric dipole of a neutron on the grounds that the measure of electric charge in the neutron was too weak to be an effective measure. In his textbook, *Nuclei and Particles,*[5] Emilio Segré pointed out that the electric dipole moment of the neutron – equal to the charge on an electron x10^{-20} – is so minute compared

4 Alvarez, L. W; Bloch, F. *A quantitative determination of the neutron magnetic moment in absolute nuclear magnetrons* Physical Review 57: 111–122. 1940.

5 Segre E., *Nuclei & Particles* Benjamin Inc (1964)

to the margin of error - 0.1 +/- 2.4 - that "this moment could be exactly zero, in agreement with the theory."

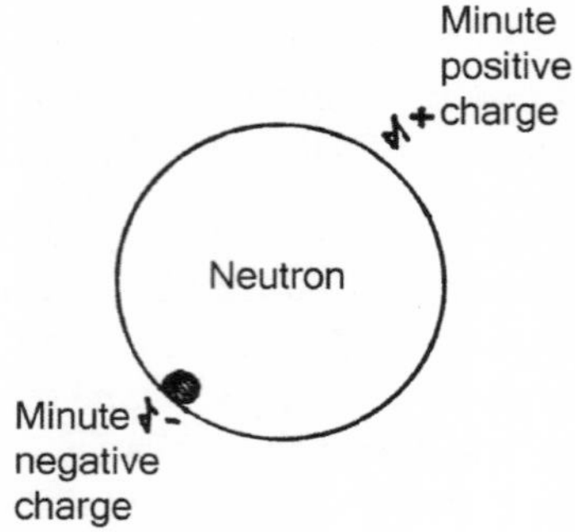

The electric dipole of a neutron would be expected to be very weak because, if the neutron were an electron bound to a proton their opposite charges would virtually cancel each other out. Despite the fact he was a world authority on neutrons, Segré, was incorrect to say the measure could be treated as zero. The margin of error, in the dipole experiment, may have indicated the measure was too weak to be conclusive but it did not indicate there was no measure. The way Segré spoke about the result of that experiment suggested that it revealed an awkward fact he wanted to hide to protect 'the theory', which could be taken to imply that the bound state model for the neutron posed a threat to a major theory in physics.

The neutron has magnetism. It has a magnetic moment of 1.91 nuclear magnetrons.[6] A particle cannot have a *magnetic moment* unless it has electric charge because the magnetic moment of a particle is created by the spin of its charge. Rather than accept the neutron must have a charge because it has magnetism, physicists ignored the evidence.

6 Alvarez, L. W; Bloch, F. *A quantitative determination of the neutron magnetic moment in absolute nuclear magnetrons* Physical Review 57: 111–122. 1940.

In high energy physics, some strange particles exit atomic nuclei with an electron seed at their centre. That suggests that there are electrons in atomic nuclei. If they are not bound in neutrons, where else in the nucleus could they be?

Chien Shiung Wu, a Chinese-American scientist, was a colleague of Segré. Known as the 'First Lady of Physics', she was a leading authority on neutrons. She was also one of the top scientists involved in the *Manhattan Project* that led to the atomic bomb. In 1956 Wu lined up the nuclei of radioactive atoms in a magnetic field so they were all spinning in the same direction and observed that more electrons were emitted in one direction than in another.

Wu's experiment revealed that when neutrons decay electrons are emitted directionally, as if they had been sitting on a specific site on neutrons from which they emerged when the neutrons decayed. This finding was more in line with the bound state theory that electrons exist as charged particles attached to protons to form neutrons than the quark theory that electrons cease to exist when they come into contact with protons to form neutrons then reappear when the neutrons decay. Also the neutrons could not have aligned in the magnetic field unless they contained charge because charged particles interact with magnetic fields whereas neutral particles do not. Physicists know this but choose to ignore it.

Physicists argue that a neutron can't be an electron bound to a proton because a neutron has the same value of quantum spin as a proton or an electron. They say that if the neutron were an electron and a proton bound together, it should display the quantum spin of them both. However, if a light electron were immobilised by a massive proton – nearly two thousand times as massive – its quantum spin would not be apparent. The greater inertia of the proton, conferred by its much greater mass, could account for the neutron appearing to have just the quantum spin of the proton, while the two particles are bound together.

Physicists also argue that the law of *Conservation of Angular Momentum* does not allow for the quantum spin of an electron to be lost in a neutron. But the neutron is unstable and physics does allow for conservation laws to apply to the *overall process of the formation and decay of unstable particles.*[7] In quantum mechanics so long as no conservation laws are broken, in the overall process of formation and decay of unstable particles, the conservation laws are considered to have been maintained. Whereas the norm in physics is to allow this flexibility regarding conservation laws within the bounds of uncertainty, the evidence pertaining to the neutron would appear to suggest that conservation laws being upheld in the overall process of the formation and decay of unstable particles may be more general.

The quantum spin of an electron appears to be conserved in the overall process of formation and decay of the unstable neutron. This is because an electron going into the formation of a neutron (in K-capture) always has the same value of quantum spin as an electron coming out of a neutron when it decays (in Beta-decay). I believe any argument in the application of a conservation law should be decided by experimental evidence not convention or arbitrary factors.

Despite all the evidence for the neutron being a bound state of proton and electron, physicists have been persistent in their determination that neutrons are not bound states of electrons and protons.

To me it was obvious physicists were attempting to hide something. Could it be they were attempting to cover up evidence that threatened one of their major theories? I could smell a rat. Or was it a black swan?

The Austrian philosopher, Karl Popper explained the scientific method of eliminating theories – in the face of incontrovertible evidence that they cannot be true – with his famous analogy of black and white swans. Someone could have a theory that all swans are white but even if a hundred white swans were counted, the theory would not be proved true and the addition of ever more white swans wouldn't make it any

7 Richards et al, *Modern University Physics,* Addison-Wesley 1973

truer. However, the appearance of a single black swan would disprove the theory altogether.[8]

No theory in science is ever considered to be absolutely true. A theory stands only as long as it isn't disproved. Karl Popper believed the role of science was more to do with disproving theories than proving them. Furthermore, the integrity of science depends on the willingness of scientists to welcome incontrovertible evidence that disproves a theory.

The gentlemen physicists of Great Britain played cricket by the rules. When the European physicist, Niels Bohr, used quantum theory to account for the discovery of *spectral lines*, the British physicists conceded that their theory of the atom being a vortex had been bowled out. They abandoned their vortex theory because it couldn't explain spectral lines. They accepted that it was the turn of quantum theory to have its innings.

The theory of quantum mechanics hinged on the uncertainty principle, developed in Copenhagen by the team under the captaincy of Bohr, was in bat for a decade until it was bowled out in 1932. Quantum mechanics should have finished its innings then. But the Europeans, unfamiliar with the rules of cricket continued to bat despite the fact the uncertainty principle had been bowled out by the neutron. That explains the persistence of the bizarre account for the neutron in physics. There has been a cover-up in physics to hide the true nature of the neutron because it disproved the uncertainty principle which is the foundation of quantum mechanics. The time has come for quantum mechanics to retire from the wicket and allow the vortex theory another innings.

8 Popper K. *The Logic of Scientific Discovery,* Hutchinson 1968

CHAPTER 4
A CHALLENGE FOR QUANTUM MECHANICS

A major pillar of quantum mechanics is the principle of indeterminacy published in 1927 by the German physicist Werner Heisenberg. Heisenberg proposed that the *Observer Effect* might apply at a quantum level. The observer effect is based on the idea that the act of observation can effect that which is being observed.

Heisenberg developed a principle to show that the process of making certain measurements, in the sub-atomic world, would increase the uncertainty about what was going on there. For example, if you wanted to look at a subatomic particle, in order to ascertain its position, you would reflect a quantum of energy off it. But the act of bouncing a quantum of energy off a subatomic particle would give it a kick, which would increase its momentum. That would make its position more uncertain.

Looking at small objects requires more energy than is required for looking at large ones. This is evident in the electron microscope, which employs higher frequency radiation than a light microscope. Because subatomic particles are the smallest things in nature, the action of ascertaining their position, with any degree of certainty, would require the use of very large amounts of energy, which would give them an enormous kick. Heisenberg argued against being able to determine with any degree of certainty both the position and the momentum of a subatomic particle.

Heisenberg's principle of quantum indeterminacy came to incorporate other features of quantum theory including *wave-particle duality* determined by Loius de Broglie and *wave indeterminacy* proposed by Erwin Schrödinger. Together they led to what is now known as the *Heisenberg Uncertainty Principle.* This principle played a central role in the development of quantum mechanics, from 1927 onwards.

The neutron, discovered at Cambridge in 1932, offered an experimental test for the uncertainty principle as it seemed obvious that a neutron must be an electron bound to a proton. In that situation both the position and momentum of the electron could be defined with certainty. The position of the electron would be somewhere within the circumference of a neutron. The momentum of the electron could not exceed that conferred on it by the energy locked up in the mass of the neutron that exceeds the sum mass of an electron and proton.

The rest mass of an electron is 0.911×10^{-30} kg. The rest mass of a proton is 1672.62×10^{-30} kg. The rest mass of a neutron is 1674.92×10^{-30} kg; therefore, it has a mass of 1.389×10^{-30} kg in excess of the sum mass of an electron and proton. This is equal to 1.5x the rest mass of an electron. In any neutron, an electron could not possess a momentum in excess of that allowed by the energy locked up in 1.389×10^{-30} kg of mass.[1]

When the uncertainty principle was applied to an electron, as if it were attached to a proton to form a neutron, the certainty in its position predicted an enormous indeterminacy in its momentum. The uncertainty principle predicted that electrons bound in a neutron could have any velocity ranging from zero to 99.97% of the velocity of light. At this high velocity electrons would possess so much energy that their mass would be in the order of forty times the mass of an electron at rest.

If neutrons were bound states of electrons and protons and the uncertainty principle were valid then some electrons in neutrons would have velocities approaching zero, while others would have velocities approaching the speed of light, and most could have some velocity in between. This wide range of velocities would be reflected in a range of mass values for neutrons from that of a single electron and proton to that of a proton and forty electrons.

The problem confronting the physicists was that all neutrons have precisely the same mass. That suggested that in the one and

1 Richards, *Modern University Physics,* Addison-Wesley 1973

only situation where the uncertainty principle could be tested it failed the test.

If the uncertainty principle were to pass the neutron test neutrons should have mass values ranging from 1673.5 x 10^{-30} kg to 1710 x 10^{-30} kg. Experimental physics, however, has revealed that all neutrons possess a very precise mass of 1674.92 x 10^{-30} kg.[2]

Physicists were faced with a stark choice. They either had to accept that the uncertainty principle was proved wrong and go back to the drawing board with quantum mechanics or reconfigure physics around the assumption that neutrons were not electrons bound to protons. They chose the latter conclusion.

By the time the neutron was discovered quantum mechanics had gained so much momentum that physicists were unable to accept its uncertainty. They decided to ignore the evidence stacking up that the neutron was an electron bound to a proton until they could find or develop a solution to the problem. The opportunity arrived with quark theory.

Using quark theory physicists were able to propose that the electron and proton ceased to exist as separate particles when they came together. By a process worthy of alchemy, they suggested that the two particles underwent transmutation into a new, entirely neutral particle. Voilá, the sticky situation of an electron having a certain position in a neutron was dismissed. The Schrödinger wave equation ensured it was impossible to be certain of an electron's position in an atomic orbit, which enabled atomic orbits to be excluded from the uncertainty principle. Now quantum mechanics coupled with quark theory would enable physicists to dispense with the nasty little neutron as well.

It is all too easy to tweak facts to fit theories. However, as Richard Feynman said, "If your theories and mathematics do not match up to the experiments then they are wrong."[2]

2 Calder N, *Key to the Universe: A Report on the New Physics*, BBC Publications 1977

The truth is that since the discovery of the neutron, physicists have been in denial of the evidence it presents that threatens the uncertainty principle. They have refused to admit to the uncertainty of their theories for forces in quantum mechanics, which depend upon the principle of uncertainty. Albert Einstein despised the emphasis placed on uncertainty in quantum theory. He knew the direction was wrong. He argued through the night with Heisenberg, on one occasion, reducing the younger man to tears. He described the uncertainty principle as, "...a real witches calculus...most ingenious, and adequately protected by its great complexity against being proved wrong."[3]

Many people will protest that the uncertainty principle must be right because of the incredible success of quantum mechanics. However, the successful application of a principle does not prove its validity. A car can work perfectly well without a certificate of roadworthiness but that does not mean it is roadworthy. A theoretical principle can work even if it is fundamentally flawed. With the failure of the uncertainty principle and the obviousness that quark theory is nonsense maybe quantum mechanics is running without a certificate of science-worthiness. Perhaps the time has come for it to be scrapped.

The problem for physicists is that the credibility of science hangs on quantum mechanics. Quantum mechanics is the most successful theory in the history of science and is the core of the advanced physics being taught in universities throughout the world. If physicists were to admit the uncertainty principle had been disproved by the neutron, they would have to admit there has been a 'cover up' of evidence because it threatened their most cherished theory. That could shatter confidence in physics and the integrity of science. Physics is the king of sciences. If physics were to fall into disrepute then the credibility of science would be threatened. For the professional scientist that is unthinkable. To quote Professor A.J. Leggett from, *The Problems of Physics:*[3]

"Quantum mechanics...has had a success which is almost impossible to exaggerate. It is the basis of just about everything we claim to understand in atomic and sub-atomic physics, most things in condensed-matter physics, and to an increasing extent much of cosmology. For the majority of practicing physicists today it is the correct description of nature, and they find it difficult to conceive that any current or future problem of physics will be solved in other than quantum mechanical terms. Yet despite all the successes, there is a persistent and, to their colleagues, sometimes irritating minority who feel that as a complete theory of the Universe, quantum mechanics has feet of clay, and indeed carries within it the seeds of its own destruction."

The neutron is the particle that drives the chain reactions in nuclear weapons. It would be divine retribution if the neutron turned out to be the nemesis of quantum mechanics. In his Inaugural Lecture entitled, *'Is the end in sight for theoretical physics?'* Stephen Hawking declared that:

...because of the Heisenberg uncertainty principle, the electron can not be at rest in the nucleus of an atom.[4]

Evidence in regard to the neutron reveals the contrary. Electrons are at rest in atomic nuclei. That is why the end is in sight for standard theoretical physics.

3 Leggett A.J. *The Problems of Physics,* Oxford University Press, 1987

4 Hawking S. *Black Holes and Baby Universes,* Bantam, 1993

CHAPTER 5
CRAZY THEORIES IN QUANTUM MECHANICS

Quantum mechanics was developed, between 1925 and 1928, at the Institute of Physics in Copenhagen, by a group of young men in their twenties who came to be known as the *knabenphysik,* 'the boy physicists'. They included, Werner Heisenberg, Wolfgang Pauli, Max Born and Paul Dirac, who worked under the wary eye of the two older men, Niels Bohr, and Erwin Schrödinger, struggling to keep up with them.

Albert Einstein, who had led the quantum revolution, turned his back on the Copenhagen initiative in quantum physics. He didn't like the notion that the quantum world was essentially unknowable. He was deeply unhappy about the way the *knabenphysik* were using uncertainty in the quantum world in the development of quantum physics.

In quantum mechanics, crazy theories were accepted as normal. In 1958, Niels Bohr, at the end of a lecture given by Wolfgang Pauli, remarked in jest, "We are all agreed that your theory is crazy. The question which divides us is whether it is crazy enough."[1]

A crazy theory that emerged from quantum mechanics was that force carrying particles could come into existence by borrowing energy from the Universe. The force carrying virtual particles were then exchanged between sub-atomic real particles to transmit the force between them after which the virtual particles vanished and repaid the energy debt. Because the appearance and disappearance of the force carrying particles

1 Calder N, *Key to the Universe: A Report on the New Physics*, BBC Publications 1977

occurred within the bounds set by the uncertainty principle, no conservations laws were deemed to have been broken.

Force carrying particles originated as mathematical constructs that worked very well within the burgeoning quantum mechanics but nobody felt obliged to believe in them. Amongst the quantum physicists in Copenhagen belief in the models of physics were denounced as *naïve realism.* Therefore, force carrying particles were called *virtual particles.*

However, experimental evidence pertaining to the neutron suggested that the theory of force carrying particles in quantum mechanics was unsound. To quote Professor Harald Fritzsch from Quarks: The Stuff of Matter.[2]

"We do not understand why the neutron is heavier than the proton. Indeed an unbiased physicist would have to assume the opposite by the following logic. It is reasonable to think that the difference in mass between the proton and neutron is related to electro-magnetic interaction since the proton has an electric field and the neutron does not. If we rob the proton of its charge, we would expect the neutron and the proton to have the same mass. The proton is therefore logically expected to be heavier than the neutron by an amount corresponding to the energy needed to create the electric field around it."

The theory of force carrying particles may have started as a useful mathematical construct to account for forces in quantum mechanics. At the onset the idea of force carrying particles may have been just a mathematical model, but then something happened to cause the physicists following of the *knabenphysik,* to think that the virtual reality in the quantum world might be real.

The 'virtual particles' became 'quantum reality' after a young Japanese physicist, Hideki Yukawa, predicted the meson, a force carrying particle capable of binding protons and neutrons in the nucleus of an atom.

2 Fritzsh H. *Quarks: The Stuff of Matter,* Allen Lane, 1983

When, in 1946, Cecil Powell discovered a shower of extremely short lived particles, knocked out of the nucleus of a silver atom, in his classic cosmic ray photograph,[3] and the particles were discovered to fit the Yukawa's predictions, it was proclaimed Powell had discovered Yukawa's mesons. He and Yukawa were awarded the 1950 Nobel Prize. From then on, in the world of physics, force carrying particles were taken to be real.

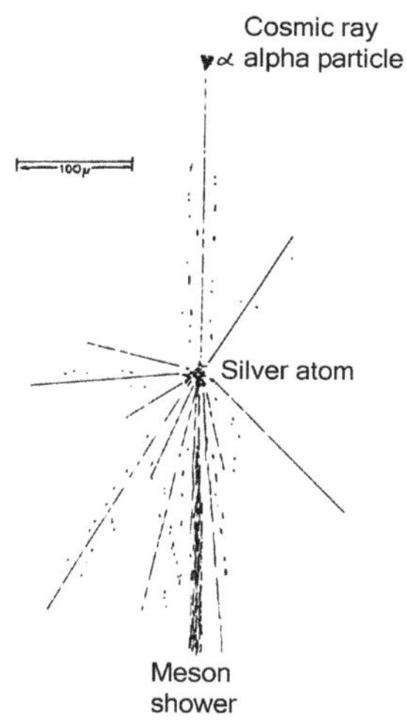

Werner Heisenberg's initial intuition of indeterminacy in the quantum world was not in itself a problem because it was brilliant and insightful and offered an opportunity to break the mould of scientific materialism. The problem came later when the principle of uncertainty gave license to the development of speculative theories, within the bounds of uncertainty, which subsequently came to be taken as certain.

3 McKenzie A. E. *A Second MKS Course in Electricity,* Cambridge University Press, plate 19, 1968

CHAPTER 6
UNDERSTANDING QUANTUM MECHANICS

Materialism, the bedrock of science, was inadvertently dispatched by Heisenberg in an initial intuition that led him to quantum mechanics. Materialism has no place in physics. Neither does the concept of a universal electromagnetic field.

The universal field, imagined as a medium to support the electromagnetic waves of light, poses a serious problem because the waves of light are *transverse.* Transverse waves involve sheer forces that require strong mediums to support them and the faster the speed of the waves the stronger and more solid the medium has to be to support them. The speed of light is the fastest speed we are aware of so any field that is supposed to carry the waves of electromagnetic radiation from the sun and stars to us would have to be the strongest and most solid medium imaginable. It is hard to envisage space filled with such a solid medium. As we cruise round the sun on a spinning planet space appears to be entirely empty. The idea that there could be a universal field more solid than steel filling space able to carry electromagnetic, transverse waves of light and heat is clearly nonsense.

We know that light exists and consists of quantum transverse waves of energy. However, experimental physics does not support the idea that there is any medium to transmit these sheer up and down waves.

The underlying problem with the classical theory of the electromagnetic field is that it embodies the world view of materialism; that *something exists to support movement.* For example, it is hard to imagine waves without an ocean.

In my early twenties, when I started this work, we lived at Crackington Haven in Cornwall and most days I would go down to the sea with my brothers to surf. We relied on the

Atlantic Ocean for the surf. We would have been bemused if we arrived on the beach one day to find the surf was still running but the ocean had gone. Heisenberg lived with that uncertainty.

Young Werner surfed the quantum wave of uncertainty. His initial flash of intuition was that the existence of particles was uncertain unless they were interacting. He treated quantum reality in terms of 'interactions' rather than 'substantial things'. The quantum mechanics Heisenberg developed, from this approach to quantum reality, proved to be incredibly successful, but I believe that success is testimony, not to uncertainty, but to the discovery of the non-materialistic nature of quantum reality.

Heisenberg intuited that at a quantum level there are no 'things' as such, there are just happenings, events and interactions. What Heisenberg established, through the mathematics he initiated, is that particles of energy could not be treated as substantial things because their existence was only certain in their interactions. This is what people found really hard to understand in quantum mechanics. As Richard Feynman, a craftsman at handling quantum mechanics, said, "I think I can safely say that nobody understands quantum mechanics."

Albert Einstein established that everything consists of particles of energy. Heisenberg established that there is no certainty that these particles of energy exist until they interact. That is the pill which people found so hard to swallow. They could think of energy as 'some thing' but they found it hard to imagine that between interactions that the 'some thing' might be 'no thing'

In classical physics energy was defined as the action of things. In the quantum revolution of Einstein and Heisenberg the thing got ditched and all we were left with was the action. The problem in physics has been the attempt to drag things into the quantum world that don't belong there. We are attached to things. We hate the idea that no-thing exists.

Heisenberg showed how particles of energy could exist in the quantum world as particles of pure action. He did this by establishing the *relational aspect of quantum reality.*

Heisenberg revealed that no single particle of energy can exist on its own; that each depends for its existence on the existence of other particles of energy to interact with. He did this by setting a rule that we only know particles of energy exist when they interact with other particles of energy. This makes sense if energy is not a thing but a particle of action.

From Newton's laws of motion we know that for every action there must be an equal and opposite reaction. So if a particle of action exists where no thing is acting, the action can only exist by virtue of its interaction with another particle of action. I believe Heisenberg intuited a key to the Universe: *Particles of energy can exist as particles of pure action only by virtue of their interactions with other particles of energy.*

This could explain why the math Heisenberg developed, with the help of the star Cambridge mathematician Paul Dirac, has been so successful in every sphere of science, engineering and technology. It is based on the great mystery of the Universe: *Action exists but no thing exists that is acting.*

I believe Heisenberg's uncovered the *Universal Law of Love,* which is that particles of energy depend for their existence on the existence of other particles of energy to interact with. The Universe operates on the principle of *love your neighbour.* Heisenberg established that the existence of particles in the quantum world can only be determined by their neighbourly interactions.

Mystics in ancient India have been aware, for millennia, that the fundamental nature of reality is the underlying dynamic state. This was epitomised by the allegory that the Universe is the *dance of Shiva.* There is no thing that dances, there is only the dance; the dance we call energy.

Yogis in India probed matter with their paranormal powers and, millennia before Einstein, they perceived how, through spin, energy can form mass. Through the yogic insight into the atom it is possible to appreciate the origin of the *myth of materialism* that blights science. The yogis actually saw particles of spin forming fundamental particles of matter and realised these 'whirlpools of light' set up the illusion of material forms. They called this Maya, the illusion of forms, which they implicated as the source of human suffering.

BOOK II

THE NEW VORTEX THEORY

Science is dead if it deals only with the material world. Material does not exist; it is an illusion set up by spin and a science that deals only with illusion is not a true science.

David Ash

CHAPTER 1
THEORIES ARE STORIES

In his book *Sapiens: A Brief History of Humankind,*[1] Yuval Noah Harari speaks of the importance of stories and myths in the success of humankind. In his own words:

"Any large scale human cooperation – whether a modern state, a medieval church, an ancient city or an archaic tribe – is rooted in common myths that exist only in people's collective imaginations...Yet none of these things exists outside the stories that people invent and tell one another. There are no gods in the Universe, no nations, no money, no human rights, no laws and no justice outside the common imagination of human beings."

Harari didn't mention the stories in science. He could have added: *There are no scientific hypotheses, axioms and theories outside the common imagination of human beings.*

Science is built on stories. The stories in science are called hypotheses and theories. The problem we all have is that we lose sight of stories being just stories. We begin to believe in our stories. After a generation or two, theories, axioms and theologies become generally accepted and then they morph into fact.

Progress is not made by believing in theories. Progress is made by challenging them; and that is why contesting theories is so important in science. Einstein and Heisenberg didn't make their breakthroughs by building on the theories that were accepted in their day. They started a revolution in physics by turning over the tables in the temples of science. They came out with their own radical view of reality, based on intuition; the momentary flash of genius. But first they had to challenge an axiom.

1 Harari Y. N. *Sapiens: A Brief History of Humankind,* Vintage, 2014

To quote Lincoln Steffans,[2]

"We know there is no absolute knowledge, that there are only theories; but we forget this. The better educated we are, the harder we believe in axioms. I asked Einstein in Berlin once how he, a trained, drilled, teaching scientist of the worst sort, a mathematician, physicist, astronomer, had been able to make his discoveries. 'How did you ever do it?' I exclaimed, and he, understanding and smiling, gave the answer: 'By challenging an axiom.'"

The standard theory in physics with its quarks and its quantum mechanics, its electromagnetic fields and its gravity theory based on a bent space-time continuum is only a story but most scientists, philosophers and educated people accept it without question. They take the story in science to be fact and forget that in essence it is a mythology. It is just a model to help us get to grips with the world we live in.

Money is an extraordinarily potent mythology. We have built an entire civilisation on it. Quantum mechanics is another. So successful are these outpourings of human imagination that we have come to believe in them with the same fervour as any Christian who believes that the Bible is the word of God.

The biggest problem in science today is that most scientists believe quantum mechanics to be infallible truth. The standard theory in physics is to most physicists what the Bible is to most Christians. But the Bible is a story and quantum mechanics is a theory; that is the infallible truth.

The stories we construct to understand things are *frames of understanding*. Each frame of understanding is like a world unto itself. An entire world of economics has grown out of the frame of money. Scientific theories set up their own frames too. The standard theory in physics is a frame of understanding which is also like a world unto itself. Physicists live in that world like Cardinals in the Vatican, and like Cardinals, they find it hard to

2 Steffans L., *Autobiography*, Harcourt Brace (N.Y.) 1931

relate to worlds in other frames. It is rare to see a Cardinal in a Hindu temple, for example. Hindus and Catholics are both into religion but for each their frame of understanding within religion is completely different.

Most physicists these days inhabit a frame of understanding dominated by quantum mechanics. They find it hard to move into another frame of understanding which does not embrace that theory. The new vortex theory is a frame of understanding built on a completely different set of axioms to those set by Werner Heisenberg. The story I write in the new vortex theory is a different story to the one written by Heisenberg and Dirac, Schrödinger and Fermi in the 1920s at the Copenhagen Institute of Physics.

Physicists will find it very hard to appreciate the new vortex theory because it does not relate to the frame of understanding in which they are deeply embedded. My grandson Fenton will find it much easier. He was reading quantum mechanics at age thirteen but because he was only thirteen he didn't become deeply embedded in that frame of understanding.

Max Planck, the father of quantum theory appreciated the problem that scientists have in changing their frames of understanding. He said: "A scientific truth does not triumph by convincing its opponents and making them see the light, but because its opponents eventually die and a new generation grows up that is familiar with it."

It would be naïve to expect the world of physics to drop quantum mechanics and embrace the new vortex theory. Quantum mechanics is a story that scientists believe in. It works as well for them as the Bible stories work for Christians. The new vortex theory is not for professors of physics, it is intended for people like Fenton.

Western civilisation is built on the civilisations of ancient Babylon and Egypt, Israel, Greece and Rome. Western religions originated in ancient Israel, our monetary systems can be traced back to ancient Babylon and science is based on the teachings of philosophers in ancient Greece; most especially the atomic hypothesis of the Greek philosopher Democritus.

Scientific materialism, hinged on the atomic hypothesis of ancient Greece, is just a story. It is a story that has been refuted by recent discoveries in high energy physics. It is a story that no longer has a place in science. Scientists who believe in materialism – and many do with the fervour of a fundamentalist – should not call themselves scientists; not when the philosophy of materialism has been disproved so frequently in cosmic and high energy research.

Democritus

We in the West have a tendency to believe that our stories are superior to the stories of other civilisations. We have a history of imposing our science and religion on other peoples with a degree of self righteous superiority. People who think they are right do not make a habit of learning from people who they believe to be wrong. So it was, when the British conquered India they did not go out of their way to learn from the Indian people. They didn't enrich Western science and religion with insights from the ancient tradition of Yoga.

Things began to change in the 1960's. A friend of my father's Barry Stonehill helped Yogi Maharishi, when he came to England from India, by accommodating him in his flat in Mayfair. Barry was rich at the time and helped finance Yogi Maharishi to do workshops around the country. One of those workshops was in Wales, in August 1967. It was attended by the Beatles and the rest is history, so to speak.

Barry knew that the Yogi Maharishi from India had something significant to share with people in the West. Barry wanted to do something to help get an all important message from the ancient tradition of Yoga out to the world. He had the cash so he helped to pay the bills.

In June 1965 I made a discovery in a book on Yogic philosophy. Like Barry I didn't do anything amazing. I wasn't a Yogi, neither was I a genius like Heisenberg or Einstein. I was just a lad who happened to be in the right place, at the right time, with the right appreciation to understand the significance of what I had discovered.

When I read the musty tome by Yogi Ramacharaka,[3] *An Advanced Course in Yogi Philosophy* which had been all but lost in an attic, in the shelves of a home library of antiquarian books, I knew immediately how important the knowledge of Yoga was. I realised that Yogis in ancient times had an insight into the atom which could save science from the error of materialism. I wanted to do something to help get that all important message, from the tradition of Yoga, out to the world. I was sixteen at the time. I had nothing to give but myself.

3 Ramacharaka Yogi, *An Advanced Course in Yogi Philosophy*, Fowler 1904, (Cosimo Classics Philosophy, 2005

CHAPTER 2
INDIA THE MOTHER OF SCIENCE

Most people in the West are unaware that science may have originated in India. In the words of James Grant Duff:

"Many advances in the sciences we consider today to have been made in Europe were in fact made in India centuries ago."[1]

Two thousand, seven hundred years ago there was a university at Takshashila in North West India, which offered 68 disciplines of learning to 10,500 students from India, Arabia, Babylonia, Greece, Syria and China. The curriculum included science and mathematics. From Takshashila the science and mathematics of India spread throughout the known world.

India invented the zero and established the decimal system, which would have been in the Takshashila University, math curriculum. This system of mathematics, essential to modern science, especially physics and information technology, is recorded in Sanskrit texts dating back to the 4th Century BCE.

There are records, as early as 100 BCE, of specific names of numbers now described as raised by ten to a power. In the west we use the word *hundred* for ten to the power of two i.e. 10^2 , and a *thousand* for ten to the power of three i.e. 10^3 Most of us stop at a *trillion* – 1000,000,000,000 – which is ten to the power of twelve i.e. 10^{12}. Modern mathematicians have names for numbers going up to 10^{30}. But in ancient India the mathematicians had names for numbers going up to the power of 10^{53} and an Indian sutra written in 100 BCE records a number raised as high as 10^{140}.

1 BAPS *Shri Swaminarayan Mandir* (Hindu temple), Neasden, London.

The decimal system was carried to the West through the Islamic empire, an empire which reached from Southern Spain to North West India. Lucid explanations of zero, translated into Arabic from Indian texts in the 7th century BCE, arrived in Spain in the 8th century of the modern era.

The nineteenth century French philosopher, Pierre-Simon Laplace, attests to the fact that:

"It was India that gave us the ingenious method of expressing all numbers by means of ten symbols, a profound and important idea which escaped the genius of Archimedes and Apollonius, two of the greatest men produced by antiquity."

Geometry also originated in India. The word *geometry* comes from the Sanskrit *gyamiti,* which means 'to measure the Earth'. Trigonometry comes from the Sanskrit *trikonamiti* meaning the measure of triangular forms. Euclid did not invent Geometry, three hundred years BCE in Greece, because records of geometry date back to 1000 BCE in India.

Many ideas credited to ancient Greece did not originate there but arrived from other lands and more ancient civilisations. The value of Pi (π) – the ratio of the circumference to the diameter of a circle – as three, approximately, appears in a 6th century BCE Sanskrit text and the ninth century CE Arabian mathematician, Ibna Musa, affirmed the value came from India. Also the theorem attributed famously to Pythagoras, was formulated much earlier by an Indian mathematician called Baudhayana.

The Indian astronomer, Aryabhateyam, declared that the Earth orbits the sun. He explained:

"Just as a person traveling on a boat feels that the trees on the bank are moving, people on the Earth feel the sun is moving."

Aryabhateyam stated that the Earth is round, rotates on its axis, orbits the sun and is suspended in space. He also explained that eclipses were the result of interplay between the sun, moon and the Earth.

Twelve hundred years before Sir Isaac Newton formulated the laws of gravity an astronomer in ancient India, Bhaskaracharya, wrote in the 'Surya Siddhanta' that:

"Objects fall to Earth due to a force of attraction by the Earth. The Earth, planets, constellations, moon and sun are held in orbit due to this attraction."

It may come as a shock to sceptics – who believe in materialism and never question the materialistic science they were taught at school – that Yogis in ancient India anticipated Albert Einstein. Using paranormal powers, Yogis probed the atom. Using a meditation technique for taking their consciousness into the atom they discovered the smallest particles of matter to be nothing but spin.

The Advanced Course in Yogi Philosophy,[2] by Yogi Ramacharaka, was published in 1904, a year before Albert Einstein published $E=mc^2$. In this book it stated that energy, known to Yogis as *prana,* exists in matter in the form of *vritta* (vortices). By a year, Yogi Ramacharaka superceded Albert Einstein, by revealing, from an ancient lost knowledge of India, not only that mass is a form of energy but that *spin* is how energy forms mass.

I discovered this thread of ancient knowledge when I was a teenager and have been following it for over half a century. In essence I discovered that light, when it spins, forms matter. With that understanding I found I could explain just about everything in physics.

Over the years all the properties attributed to substantial material particles, such as mass, inertia, three dimensional extension and fields of force, as well as nuclear energy, space, time, gravity and dark energy, I found I could account for with the vortex theory originating in the mysticism of ancient India. Yogis recognised that what we in the West perceive as material substance is an illusion set up by spin. The Indian mystics called this illusion *Maya,* the illusion of *forms.* Now I realise materialism is the ultimate spin of Western civilisation. People who believe in materialism are seemingly deluded.

2 Ramacharaka Yogi, *An Advanced Course in Yogi Philosophy,* Fowler, 1904 (A facsimile is available in Cosimo Classics.)

I have always had a passionate belief in science but the problem, as far as I can see, is that science is still anchored in the materialistic philosophy of ancient Greece. Over the last fifty years I have worked to shift the anchor of science from the philosophy of ancient Greece to the Yogic philosophy of ancient India, which is more in line with quantum theory than the Greek philosophy of materialism, which is destroying our world.

CHAPTER 3
VORTEX PARTICLES

Yogis in the pre-scientific era probed matter with paranormal powers called *siddhis*. This practice by Yogis was recorded about 400 B.C.E. in the 'Yoga Sutras of Patanjali'[1] where the results of meditation were described in detail. In Aphorism 3.46 it states that through meditation the Yogi can gain an extended faculty of observation from the practice of the *anima siddhi* – the ability to shrink consciousness commensurate with the very small.

Mystics in the ancient Indian tradition of Yoga observed subatomic matter in meditation. Using the anima siddhi Yogis perceived vortices of energy in the atom. Is it possible they actually saw *quantum spin*? Did they observe quantum reality, in contravention of the most fundamental principle in quantum mechanics?

According to the Heisenberg *Uncertainty Principle,* we are not supposed to be able to look into the quantum world. But the Yogis were not looking into the atom in a mechanical way. They were not bombarding atoms in a massive *atom smashing machine*. They made their observations through subtle manipulations of consciousness that we in the West have yet to comprehend. Their way was not to conquer the atom by bashing it with high energy particles but to conquer the mind through meditation and train it to penetrate the atom by pure consciousness in order to seek out its secrets.

If the Yogic perception is correct then subatomic particles, as vortices of energy, would be spheres without apparent poles. This is because the *degrees of freedom* for energy spinning freely

1 Feuerstein G. *The Yoga-Sûtra of Patanjali*: Inner Traditions International; Rochester, Vermont, 1989

on a single point suggest it would form a *ball vortex.* This principle operates with a ball of wool. The degree of freedom for wool to wind in every direction on infinite axes, results in the formation of a ball, a ball vortex of wool.

Balls of wool can help us understand why subatomic particles, as vortices of energy, would appear to be corpuscles without poles. The ball of wool is a vortex but it has no poles because, in the ball, the wool spirals on constantly changing axes. In much the same way the spin of energy in the subatomic vortex would also be on constantly changing axes. Therefore in the subatomic vortex of energy, as the axis of spin would be constantly shifting, its poles could not be discerned.

Energy spinning freely to form a subatomic particle of matter would form a *spherical vortex of energy.* This would explain why subatomic particles appear as corpuscles.

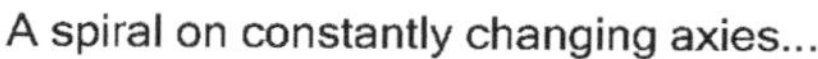

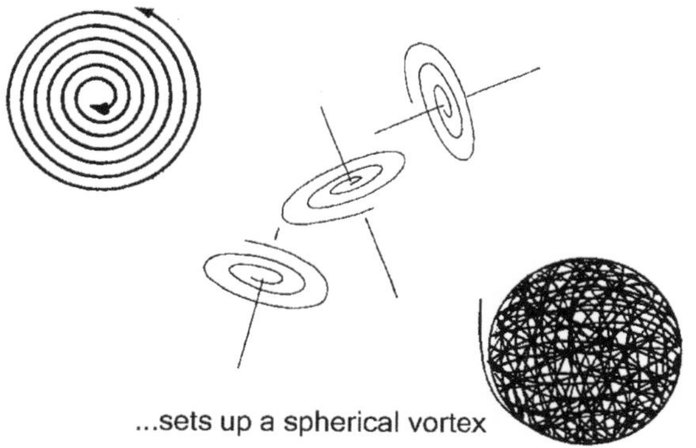

The wool ball is a compact form of wool. Likewise the spherical or ball vortex would be a very compact form of energy. If a ball of wool is unraveled it can release a lot of wool. Unravel nuclear vortices and the energy released, according to $E=mc^2$, would be enough to destroy an entire city in a single explosion.

In the vortex theory mass is defined as quantity of vortex energy. The subatomic vortex stores energy as mass. Vast amounts of energy are stored in the mass of an atom. The vortex also stores this *potential energy* in forces, like gravity, magnetism and electric charge, associated with vortex interactions.

The vortex theory is a version of string theory. Most string theories attempt to account for everything in terms of 'strings of vibration'. The vortex theory introduces the idea that subatomic particles are 'spinning strings'. A particle of energy is not literally a spin of string. It is just spin. String is a model to help us imagine particles of energy as 'lines of the movement of light' either undulating in a quantum of light or spinning in a subatomic particle of matter. The lines do not exist as physical lines. They are just imaginary lines.

The vortex is a three dimensional spiral. 3D subatomic vortices of energy, as particles of matter, would confer three dimensions on matter. We take the 3D extension of matter for granted. The vortex of energy reveals why our world is 3D.

Material substance is defined in terms of inertia, mass and three dimensional extensions. These are properties of the vortex of energy. All properties of material substance can be accounted for in terms of the 3D spin of energy in the subatomic vortex.

The *static inertia* of matter – its tendency to stay put unless forced to move – could be conferred upon it, not by material substance but by spin in the vortex. Spin creates inertia. This is illustrated by gyroscopes and fidget spinners. The spin of a fidget spinner or a gyroscope resists motion out of the plane of spin, as does the spin of a pebble on a pond. The spin of the pebble prevents the pebble falling out of the plane of spin, which is why it can skip over the surface of the water.

The spin of energy on infinite planes to form a ball vortex explains the resistance of massive subatomic particles – like protons – to movement in any direction. Static inertia is conferred upon the atoms, molecules and crystals in bodies of matter formed out of these minute whirlpools of light. Explaining the origin of inertia in matter is an achievement because in *The Character of Physical Law*[2] Richard Feynman said, "The laws of inertia have no known origin."

2 Feynman R, *The Character of Physical Law*, BBC Publications, 1965

The interactions of particles of energy can best be understood in terms of inertia. Motion sets up inertia. For example, it is easier to balance on a bicycle when it is moving than when it is standing still. This is because of the inertia conferred on it by its motion.

Propagating wave particles of energy – the basis of light – set up *kinetic inertia.* Their inertia is to keep moving unless they are stopped. Static vortex particles of energy – the basis of matter – set up *static inertia.* Their inertia is to stay put unless they are forced into motion.

The model of the ball vortex shows clearly how the illusion of materiality is set up by spin. The Western philosophy of materialism has been used to explain away mysticism. The insight of the vortex of energy, originating from the mystical tradition of India, can be used to explain away materialism.

Why do we still believe in materialism? Even in Victorian physics the material *billiard ball* model for the atom was dismissed in favour of a vortex model. The original vortex theory was that the atom was not a material particle but a vortex. The idea of the *vortex atom* was proposed by the German physician and physicist, Herman Helmholtz and championed by Lord Kelvin (William Thomson) who was the leading figure in British physics in the second half of the 19th Century.

In Kelvin's day the atom was thought to be the smallest particle of matter but Kelvin despised the common assumption that the smallest particles of matter were solid material particles like billiard balls. To him this model was unsatisfactory as it offered no explanation for the properties of matter. He considered the popular, materialistic view of matter to be superficial and naïve. He dismissed the billiard ball atom as:

"...the monstrous assumption of infinitely strong and infinitely rigid pieces of matter...Lucretius' atom does not explain any of the properties of matter..."[3]

3 Thomson W. *Popular Lectures and Addresses* 1841

With the endorsement of Lord Kelvin the theory of the vortex atom dominated physics in Victorian England in the latter half of the 19th Century. It was taught at Cambridge until 1910.

James Clerk Maxwell, who developed the equations for electromagnetic theory, was a strong proponent of the vortex atom. In the Encyclopedia Britannica of 1875 he wrote:

"...the vortex ring of Helmholtz, imagined as the true form of the atom by Thomson (Lord Kelvin), satisfies more of the conditions than any atom hitherto imagined..."

J. J. Thomson (1856-1940), who discovered the electron, was professor of physics at Cambridge when he said:

"…the vortex theory for matter is of a much more fundamental character than the ordinary solid particle theory."[4]

With the vortex theory physicists were able to reduce the properties of matter to the dynamic principle of vortex motion. But the Victorian vortex theory failed because it was applied to the atom and not to subatomic particles. That was not the fault of the Victorian scientists. They had no idea that in the 20th century to come that the atom would be split.

The theory of the vortex atom was abandoned because it was unable to explain the lines that appeared in the light spectra of atoms. The Danish physicist, Niels Bohr, explained spectral lines by applying quantum theory to the atom. He suggested the atom was a system of electrons, in distinct orbits, that could move into higher orbits. Spectral lines, Bohr explained, represented the energy signature of an atom as its electrons absorbed energy to make quantum leaps from lower to higher orbits, or emitted energy as they fell back to their original quantum states.

Werner Heisenberg took a quantum leap in thinking to suggest it was only possible to be certain the electron existed when it absorbed or emitted light but during the quantum leap maybe it didn't exist at all. With the help of a brilliant Cambridge

4 Thomson J.J. *Treatise on the Motion of Vortex Rings*, Cambridge University 1884

mathematician, Paul Dirac, the enormously successful *matrix math* of quantum mechanics came into being simply by treating particles more as relational interactions than substantial things. Whereas people found it hard to comprehend quantum reality purely in terms of action and interaction, the die was set by cosmic ray research.

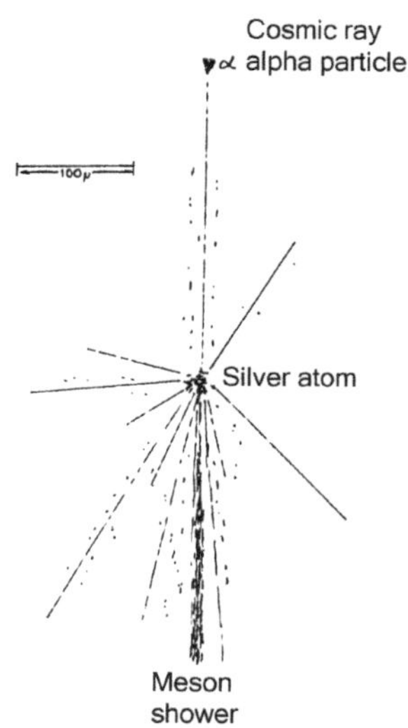

In 1946, a photograph from the cosmic ray research of Cecil Powell, professor of physics at Bristol University, witnessed a high energy cosmic ray particle slamming into the nucleus of a silver atom in a photographic emulsion.[5] When the cosmic particle was stopped by the nucleus, its kinetic energy – its

5 McKenzie A. E. *A Second MKS Course in Electricity,* Cambridge University Press, plate 19, 1968

movement – continued on, to be transformed into a new particles of matter called *pi-mesons.*

Powell's cosmic photograph witnessed transformation of kinetic energy, or movement, into mass. It provided firm experimental proof that the Universe is made of movement where there is no underlying material that moves. This discovery confirmed Heisenberg's hunch that in quantum mechanics particles could be treated purely as interactions rather than things interacting. The non-material nature of quantum reality has been repeated in high energy labs, like CERN, that came in the wake of cosmic ray research. This experiment provided irrefutable evidence that materialism is unscientific; it should be thrown out of the body of science.

Imagine rain falling to form streams tumbling down mountain sides. These join the torrent of rivers that pass through hydroelectric plants where the fall of the water is converted into the spin of turbines and then the flow of electricity. The electricity is then fed into CERN where it is used to accelerate protons. As the protons collide, in the intersecting rings of the particle accelerator, their motion is arrested. The kinetic motion originating from flowing water has been transformed into the mass of the new particles. Would anyone suggest that in the fall of rain and the tumble of streams, the torrent of rivers and the spin of turbines, the flow of electricity and the acceleration of protons, some cumbersome material substance was transferred, 'clinkety clank' and transformed, 'clomp' into the new particles? Motion alone was transferred between each step of the process. As nothing went into the newly formed particles of matter but motion, it is obvious that they can be nothing but forms of motion. Materialism is dead. It is nonsense.

Materialism should have been abandoned as naïve realism but instead the vortex model was abandoned – along with all the other classical models of physics, apart from lines of force. That was a pity because the vortex model, applied at a subatomic level, helps us understand how kinetic energy can be transformed into mass. If *mass* is treated as *quantity of vortex energy* then after a nuclear collision, in a high energy lab like CERN, the kinetic energy forced through vortices in the nucleus of an atom could be transformed into mass simply by being forced into vortex motion. This would occur as they pass through the nuclear vortices. The transformation of energy into mass could be understood to be nothing more than the transformation of one form of motion into another. High energy physics could be treated as the study of transformation of the wave form of *kinetic motion* into the vortex form of *static motion* – a system of motion that stays in one place due to spin, like a cat chasing its tail.

In the new vortex theory, proton particles in the nucleus of an atom are treated as stable, natural vortices of energy. Kinetic energy forced through them in high energy physics is transformed into short lived, synthetic vortices. These are unstable because they are unnatural. These energy-mass transformations could be imagined as batter being forced through a doughnut mold. The batter comes out of the doughnut mold in the shape of a doughnut. If it falls to the floor, in less than a second it reverts to a mess of batter. This simple analogy explains what happens in high energy physics.

CHAPTER 4
HIGH ENERGY PHYSICS

The vortex theory is the antithesis of materialism and is more believable than quark theory. The vortex theory is an improvement on quark theory as it can account for the disparity in lifespan between short lived synthetic particles and natural protons.

The naturally occurring proton is very stable. This fact is self evident from the stability of the Universe. According to the vortex theory the short lived vortex particles, appearing in cosmic ray and accelerator experiments, are generated by forcing vast amounts of energy through natural proton vortices. According to the vortex theory, natural vortex particles, like protons in atomic nuclei, can act as vortex dies.

As a boy, in my grandfather's workshop, I remember turning a rod of steel round and round into a spiral die and watching as it came out the other side in the shape of a screw. That experience prepared my young mind for the way I could use the vortex theory to explain how new particles of matter are synthesized in CERN and other high energy laboratories.

In the vortex theory I explained that the particles generated in the experiments were short lived because the vortex was not their natural form of energy. Particles of energy in the wave form only assumed the vortex form if, in a high energy experiment, they were forced through an atomic nucleus. As soon as they escaped from the nucleus, the unnatural vortices unraveled and then, reverting to their natural wave form, they radiated away as heat or light.

I used the doughnut analogy to explain this point. Batter takes on the shape of a doughnut when it is forced through a doughnut mould but if a vat of boiling oil is not placed under the dropping doughnuts, in less than a second, after leaving the mould, the doughnuts would cease to be doughnuts and revert to a mess of batter.

The more energy the physicists used in their experiments, the more massive the 'doughnut' particles they produced. That, I explained, is why they kept discovering heavier particles as they probed atomic nuclei with ever increasing amounts of energy.

In the high energy laboratories, physicists discovered that some of the new particles lasted a lot longer than others. Though the difference in life span was between a trillionth and a billionth of a second, it was significant enough for the author of quark theory, Murray Gell-Mann, to call them *Strange*.

I found I could explain the longevity of strange particles with my vortex theory. When strange particles decay they leave a proton or an electron behind. This suggested to me that when energy is blasted through an atomic nucleus, in a high energy experiment, some of it swirls around an electron or proton already occurring in the nucleus.

This swirling of energy around a natural vortex particle could explain how a *strange particle* forms. I imagined a strange particle forming, like a hailstone in a cloud, as ice crystalises, around dust particle.

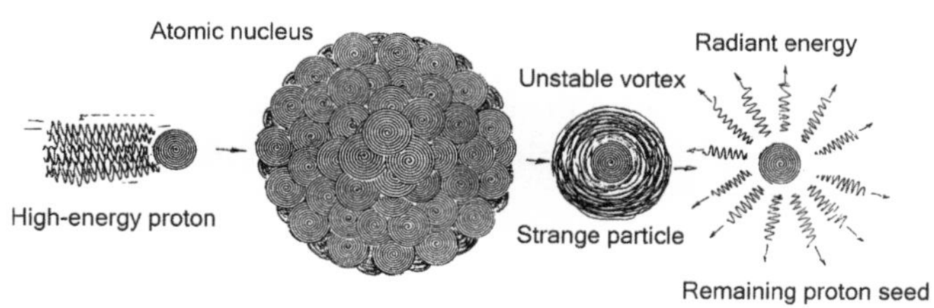

Imagine energy swirling around an electron or a proton 'natural vortex of energy' as it blasts through the nucleus of an atom. The stable vortex particle, at the centre of the maelstrom of energy could stabilise it. On exiting the nucleus the newly formed swirl of vortex energy, with the stabilising natural vortex at its centre, could last longer – that is it could take longer to unravel – than a swirl of energy that doesn't have a *scaffolding* natural vortex at its centre. As physicists increased the energy in their accelerators they found progressively more massive particles were generated in layers around a natural particle; in the same way that ice forms in layers around a particle of dust in a hailstone.

The massive particles decay in steps, or layers, to reveal a new, lighter particle, at each lower level of energy where particles tend to appear. This *cascade decay* of unstable particles fits with the model of a hailstone melting layer by layer, in reverse of the way it formed. The levels of mass-energy, forming round a proton in the creation of a strange particle, are called *doses of strangeness.* The decay of the strange particles, in steps, to reveal lighter particles on each step, is described as shedding doses of strangeness.[1]

According to the vortex theory, the stable proton vortex, at the core of the unstable strange particle, exerts a holding influence over this *transitional swirl* of energy – which I call *transitional mass*. As each dose of strangeness is shed, to leave a smaller, less massive particle behind, the stabilising influence of the scaffolding proton vortex over the residual transitional mass would be stronger. This is because there is less swirling energy to stabilise and, at the same time, the unstable swirling vortex energy would be closer to the stabilising proton vortex at the centre of the swirl, which would increase the strength of the scaffolding effect.

According to the vortex model, as each dose of strangeness is shed, the particle left behind should last longer than the one that went before it. This is evident in the cascade decay of a particle called the *Omega-minus*.

In February 1964, at the Brookhaven National Laboratory in America, a group of physicists, led by Nicholas Samios, photographed a high energy experiment that witnessed the cascade decay of a new heavy particle which was called the omega minus.[1] If you look closely at the drawing I made from the photograph of the three step cascade decay of the omega minus you will notice that the distance between each successive step, in the progressive cascade decay, is longer. The length of the track a

1 Calder N, *Key to the Universe: A Report on the New Physics*, BBC Publications 1977

particle leaves before decaying gives a measure of its life span. This shows that each residual particle, in the cascade decay process, lasted longer than the particle that went before it.

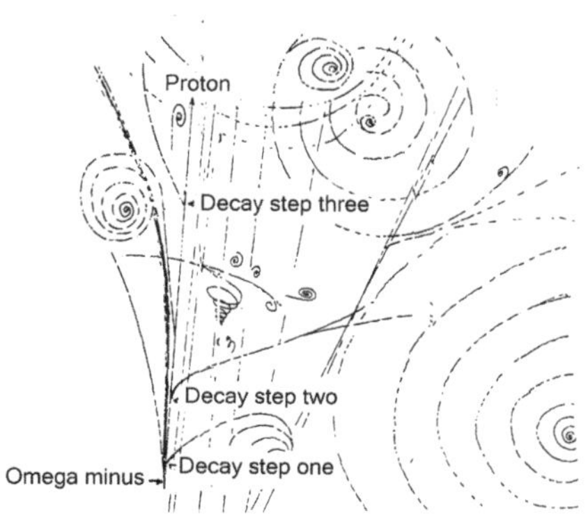

The omega minus had been predicted by Gell-Mann, and its decay in three steps, shedding a dose of strangeness at each step, was hailed as a triumph for his quark theory. However, the cascade decay of the omega-minus is also a triumph for the vortex theory. This shows how the same experiment can support two totally contradictory theories.

Physicists talk about discovering new particles of matter in their high-energy research, but to say the new particles had been discovered is misleading, because discovery implies that something already exists and is waiting to be found. According to the vortex theory the new heavy particles, in the particle zoo, are unnatural. They do not exist in nature. They have been synthesized in the accelerators. They had no place in normal matter. They are the products, not the discoveries, of high-energy physics.

High energy laboratories, like *Fermilab* in the USA or CERN in Europe resemble the Vatican with a basilica like building crowning a circular ring. Priests in religion congregate in basilicas and cathedrals to substantiate their implacable faith in monotheism. Physicists in science congregate in high energy labs to substantiate their implacable faith in materialism. Like God in monotheism, the quark in materialism has never been seen but the faithful still believe they will find the object of their fervour one day.

Scientists claim the existence of quarks has been confirmed by evidence from bombarding protons with high speed electrons in experiments – as at the Stanford Linear Accelerator (SLAC) in the 1960's. In these experiments electrons ricocheted as though they hit something hard inside the protons. Physicists concluded the hard things were quarks because quarks were what they were looking for. However, these experiments did not prove the existence of quarks, they merely showed there was something hard in the proton, which may have been something other than a quark.

CHAPTER 5
NUCLEAR ENERGY

In the vortex theory protons are considered to have a 'hard heart' because they contain *captured energy.* According to the vortex theory the SLAC electrons may have bounced back off captured energy in the protons. In the vortex theory the captured energy in the proton constitutes the mass of the primary *pi-meson*.

According to the vortex theory, over the vast periods of time protons have been is existence they have captured energy. The captured energy comes from quantum particles of propagating energy that have become trapped from moving at the speed of light, down into the proton's progressively tight spiral. By virtue of its spiral shape, a vortex of energy could trap wave energy and transform it into mass.

Living on the coast of Cornwall I was surrounded by lobster pots so I visualised the proton vortex as a lobster pot and the propagating quanta as lobsters. Lobsters crawl into a sedentary lobster pot and are caught by its cunning shape. In like manner a quantum driving into a proton vortex could become trapped in its internal spiral shape. The energy trapped in a proton could constitute the *pi-mesons* discovered in Cecil Powell's cosmic ray experiment. In that experiment one could imagine the cosmic ray particle hitting the 'lobster pot' protons, in the nucleus of the silver atom, so hard that it knocked the captured meson 'lobsters' right out of their 'vortex' pots.

If the meson energy, swirling inside a proton, is the equivalent of a lobster caught in a lobster pot, someone looking for crabs could shoot into the pot and conclude from the ricochet of their projectile that there was a hard-shell crab in the pot when it was actually a hard-shell lobster. So, when shooting electrons into protons at SLAC, the physicists hitting something hard may have thought that they had discovered quarks. However, it may not have been quarks that they discovered. They may have been hitting mesons.

To appreciate the structure of the proton vortex, imagine it as candy floss at a fair. Now think of the captured meson energy as additional strands of sugar that have found their way into the space in the spun sugar vortex. They fill it so completely that the candy floss ends up more like a hard candy ball (we called these *gobstoppers* when I was a boy) than fluffy floss. In like manner, proton vortices could appear as dense corpuscles, with increasing hardness toward their centres. This increasingly dense hardness inside them may be caused by the captured energy swirling in them. In the vortex theory the diameter of the proton is thought to be defined by the limitation of its ability to *completely capture* quantum particles of energy. That would be a feature of the tightness of its three dimensional spiral.

In my vortex theory I imagined proton and neutron vortices to be saturated with energy they had captured. I considered captured energy to be responsible for the meson mass that they contained. The high energy cosmic ray particle smashing into the nucleus of a silver atom, in Powell's experiment, released the swirling mass of meson energy from the nuclear vortices.

I also likened the nucleus of an atom to a prison. If the walls of a prison are destroyed in a bombardment prisoners can escape. As soon as they do so they cease to be prisoners. Scrambling over the demolished walls they revert to the state of free running people, as they were before they were captured. Something like that happened to the meson energy.

Analysis of Powell's photographic plate showed that the mesons, knocked out of the nucleus of the silver atom, lost their mass in the time that it takes for a quantum of energy, traveling at the speed of light, to traverse an atomic nucleus. The time amounted to approximately one hundred trillionth of a second (10^{-15} Sec).

It seemed to me that the kinetic energy transformed into mass as it passed through the atomic nucleus. As soon as the kinetic energy escaped the nucleus it reverted immediately to the wave form again. I used Newton's *Laws of Motion* at a quantum level to account for these transformations of energy into mass. I called them the *Quantum Laws of Motion.*

I was first introduced to Newton's Laws of Motion when I studied for my O level in physics at the Hastings College of Further Education. There, during my first week on the A Level physics course, I was given a textbook[1] in which I saw Cecil Powell's Nobel Prize winning cosmic ray photograph. My O level lessons were still fresh in my mind. I had also only recently discovered the vortex in Yogic Philosophy.

As I stared at the picture, in the back of my textbook, of the cosmic ray particle smashing into the atomic nucleus I realised that motion could be transformed into mass. I imagined vortex particles in the nucleus could be responsible for the transformation. It also seemed obvious to me that Newton's Laws of Motion must apply to the situation if it was motion that was being transformed into mass. A decade was to pass before I formulated my quantum laws of motion but the seeds had been planted in my mind when I was eighteen.

My first quantum law of motion stated that: **Particles of energy will maintain their form of motion unless a change of form is forced upon them.**

My second quantum law of motion stated that: **If the force of change is removed the particle of energy will revert to its original form of motion.**

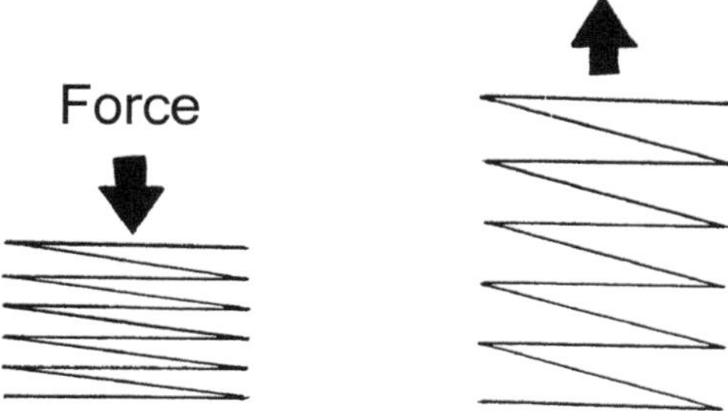

My thinking behind the quantum laws of motion was that particles of energy were like springs. A spring can be deformed

1 McKenzie A. E. *A Second MKS Course in Electricity*, Cambridge University Press 1968

by the application of a force but if the force is removed the spring will revert to its original form.

The prisoner analogy also helped me explain the quantum laws of motion. The first quantum law of motion suggested that when quantum felons are captured and imprisoned in vortex prisons they lose their freedom and contribute to the mass of the prison. The second quantum law of motion inferred that when the quantum felons escape from the vortex prison they immediately regain their freedom. They cease to be prisoners and no longer contribute to the mass of the prison. They become quantum ex-prisoners in the full flight of freedom.

While imprisoned the quantum prisoners add their mass to the mass of the prison. When they escape the prison the quantum ex-prisoners subtract their mass from the mass of the prison.

In the vortex theory mass is defined as quantity of vortex energy. The captured quanta only have mass while they are imprisoned in the vortex and are forced to spin. When they escape they do not have mass anymore because they cease to spin in the vortex and revert to their original wave form.

In accordance with the first quantum law of motion, quantum waves of energy, when captured by a vortex of energy, are forced into vortex motion. That is when they are transformed into mass. If they escape the stable vortex they will revert immediately from spin to their original wave form and lose their mass in accordance with the second quantum law of motion.

In the high temperatures of the sun, or an exploding hydrogen bomb, protons and neutrons have sufficient kinetic energy to attain very high velocities and collide with an immense force. When they do so they may converge. If proton or neutron vortices collide and converge their capacity to contain captured energy would be reduced so some of the captured energy, the *meson mass* swirling inside them, would be lost as radiant energy.

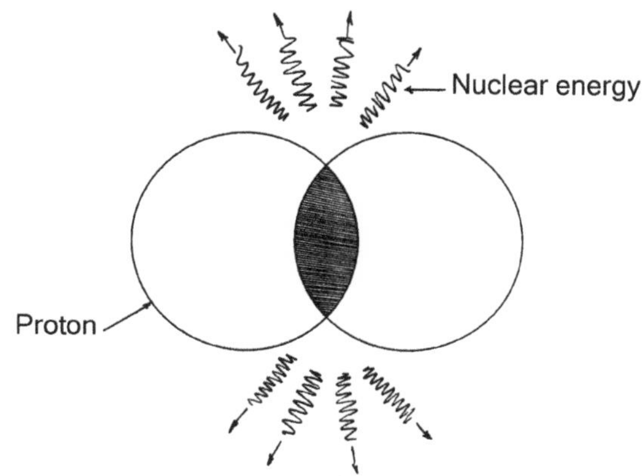

In accordance with the second quantum law of motion the swirling meson energy, escaping from the proton vortex, would revert to the original wave-propagating form and radiate away. This is how, in the vortex theory, I accounted for nuclear energy.

Protons and neutrons, saturated with captured energy, could be likened to buckets full of water. If one bucket of water is placed inside the other its capacity to contain water would be reduced so some of the water would spill out. In like manner as two protons converge their capacity to contain captured energy would be reduced so some of the captured energy would spill out. That is as nuclear energy.

Captured vortex energy constitutes part of the mass of a proton or neutron, the loss of any of this captured energy would be associated with a loss of the overall mass of the proton or neutron. In nuclear fusion 0.7% of the mass of the combining protons or neutrons is lost. This loss of mass, which is a loss of captured energy, accounts for the release of nuclear energy that occurs as hydrogen nuclei fuse together to form helium in the sun or in a hydrogen bomb.

The vortex theory can account for nuclear energy whereas, according to Richard Feynman, standard theoretical physics cannot:

"Nuclear energy…we have the formulas for that, but we do not have the fundamental laws. We know it's not electrical, not gravitational and not purely chemical, but we do not know what it is."[2]

In *Hidden Journey*[3] Andrew Harvey quoted an Oxford academic as saying, "Only scientific criteria for truth are valuable and mystics are pathological cases."

Mystics, in the pre-scientific era, probed the atom with supernormal powers and elucidated the holy grail of science; the single principle that explains everything, including nuclear energy. In the light of this extraordinary achievement maybe the time has come for scientists, philosophers and sceptics to review their attitudes towards mystics.

Mystics are not pathological cases. It would seem more likely that physicists who invent nuclear weapons of mass destruction are pathological cases. As their scientific theories lie in ruins, university academics could meditate on a very real possibility that the mystical insights of Yogis may become the basis of a new scientific criterion for truth.

2 Calder N, *Key to the Universe: A Report on the New Physics*, BBC Publications 1977

3 Harvey A. *Hidden Journey* Rider, 1991

CHAPTER 6
NUCLEAR BINDING

The *strong nuclear force,* which binds protons and neutrons together, in the nucleus of an atom, is different to electric and magnetic forces in that it does not have polarity neither does it have infinite extension. It is very short ranged. The nuclear force is limited to the diameter of the atomic nucleus.

The nuclear binding of the protons and neutrons, in the helium nucleus, which results from the nuclear fusion, can be explained with the vortex theory. As protons and neutrons converge most of their captured energy would be still be left inside them. This residual captured energy could then begin to swirl between them. The captured vortex energy, swirling between two or more converged nuclear vortices could be effective in fusing them together.

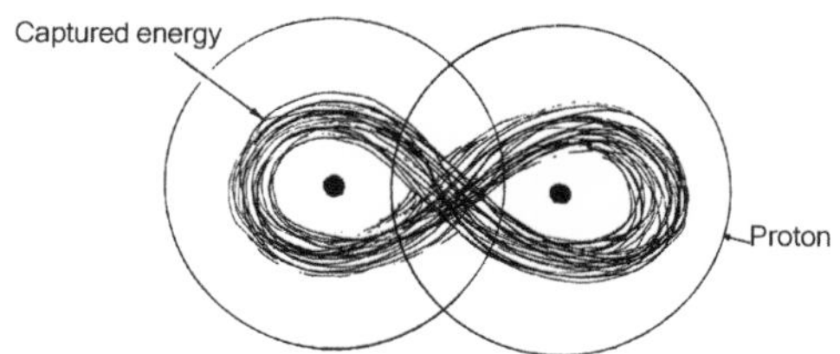

In the vortex theory I use the idea that captured energy can be *shared* between natural vortices to account for the binding of protons and neutrons in the nucleus of an atom. It is the meson mass – the captured vortex energy – swirling around two vortex particle centres, which may be responsible for the *nuclear binding* between them.

The vortex theory can also explain why the binding between nuclear particles increases when the binding mass decreases. This is currently a mystery in physics.

After nuclear fission – which occurs in the atomic bomb and nuclear reactors – the vortex particles in the nucleus converge closer. This happens after the fission of uranium or plutonium. Uranium and plutonium have large nuclei. When these divide they split into smaller nuclei. In the smaller nuclei the nuclear particles – protons and neutrons – are bound more tightly together. In the smaller fragment nuclei, the distances between the centers of the nuclear particles are less than they are in the large uranium or plutonium nuclei.

According to the vortex theory, after fission of the uranium or plutonium nuclei, the nuclear vortex particles in the two smaller nuclei clamp together more tightly. As they do so a small amount of the captured meson energy is squeezed out. This is released as nuclear energy. Meanwhile the majority of the captured energy is left swirling between the nuclear particles. The tighter binding would be caused by the residual captured energy, in the smaller nuclei, racing in a shorter circuit. After nuclear fission, the loss of a small amount of binding energy would be marginal compared to the increase in binding between nuclear particles due to the large amount of residual captured energy racing in a much shorter, tighter circuit between them.

CHAPTER 7 THE HEDGEHOG ANALOGY

In *The Vortex Theory*[1] I used a 'Hedgehog Analogy' to account for nuclear energy and nuclear binding. Imagine the proton as a hedgehog and the captured energy as its fleas. In much the same way that a hedgehog carries a resident population of fleas so a proton carries a population of captured quantum particles of energy.

Hedgehogs don't do anything to acquire fleas; the fleas just hop onto them. So it is with a proton, it does nothing to acquire its captured energy; the quantum wave-trains of kinetic energy drive into the spiral space of the proton vortex.

If a hedgehog is weighed, the greater part of the mass would be that of the hog. A lesser part would be that of the fleas. In like manner the greater part of the measured mass of a proton would be that of the stable proton vortex, but a lesser part would be that of the unstable swirl of captured energy it contains.

1 Ash D. *The Vortex Theory*, Kima Global Publishing, 2015

The hedgehog prickles represent the charge repulsion between protons. Because of their prickles hedgehogs converge with difficulty. So it is because of their charge repulsion protons that converge only do so if they collide with force.

Imagine if two hedgehogs were pushed together, as their prickles converged there would be less space between them for fleas and so some of the fleas would be evicted and hop away. In like manner, as two protons converge there would be less space within them for captured energy and so some would be lost and would radiate away as nuclear energy.

Because of the loss of some of their fleas, the weight of the converged hedgehogs would be marginally less than the sum of their weights before they were pushed together. So it is the mass of two converged protons would be less than the sum of their masses before they collided.

Most of the fleas would remain on the hedgehogs and not being bothered about which back they bite, the fleas would hop from hog to hog. In like manner captured energy belonging to the proton vortices, unconcerned within which vortex it swirls, would circulate between the converged protons and by this action would bind them together.

CHAPTER 8
ELECTRIC CHARGE AND MAGNETISM

The binding force between nuclear vortices, caused by captured energy swirling between them, would be in the central spiral region where the vortex is tight enough to hold energy in capture. Beyond this tight central region the proton vortex energy would extend into infinity. In the vortex theory the interactions of extending vortex energy are considered to be the cause of the forces of electric charge and magnetism.

The infinite extension of countless proton vortices could be responsible for the majority of positive charges in the Universe. The majority of negative charges in the Universe would be caused by an equal number of electron vortices extending into infinity. The equal numbers of opposite charges cancel each other out. This could explain why space is electrically neutral.

If charge is set up by the extension of a vortex of energy it could not be divided. This is born out by experiment. Every charged particle has the same value of unitary charge which is not affected by its mass. Imagine charged particles as soldiers. Some soldiers are big and others are little but each is armed with a rifle and so every soldier, regardless of his mass, is equally effective at a distance.

The quark army is weird. There isn't a single soldier with a whole rifle. Each quark, with its fractional charge, is like a soldier with a bit of a rifle. Three soldiers have to be together for an operative weapon. That is one of the reasons why quarks are unlikely to withstand the onslaught of vortices – each armed with unitary charge like a complete rifle – as the new vortex theory takes its stand against the quark theory.

The mass of a vortex is concentrated at its centre whereas charge is the effect of the dynamic nature of the vortex at a distance, as it extends into infinity. This explains the ability of charged particles to act-at-a-distance. In the vortex theory electric charge is not something existing separate from a subatomic particle of matter. Electric charge is considered to be an effect of the extension of the vortex particle into infinity, carrying with it the innate dynamic nature of the vortex of energy.

The vortex theory provides an account for forces that doesn't require the uncertainty principle or the assumption of virtual particles of electromagnetism. If protons and electrons are infinitely extending vortices of energy, then the electric and magnetic forces between them could be accounted for in terms of overlapping and interacting fields of real, dynamic vortex energy. These electric and magnetic interactions are detailed in *The Vortex Theory*.[1]

The quantum physicist, Richard Feynman invented *Quantum Electro Dynamics* as a theory for electromagnetic forces, based on quantum mechanics. His theory of quantum *electrodynamics,* developed at Cornell University, Ithaca, NY, was that 'virtual particles' of light continually pop up alongside charged particles of matter. The exchange of these *virtual photons* with other charged particles set up the electromagnetic force fields between them. Feynman drew simple diagrams to illustrate his QED.

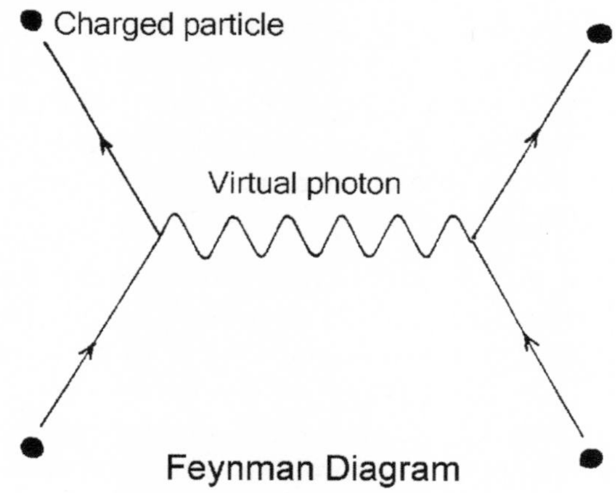

Feynman Diagram

1 Ash D. *The Vortex Theory,* Kima Global Publishers, 2015

So long as the virtual particles of light appeared and disappeared within the bounds of Heisenberg's uncertainty formula no one could say for certain they didn't exist. Feynman's QED may work well mathematically and be supported by experiments but is full of arbitrary assumptions.

The theories of quantum mechanics, such as QED, may work well but they are not easy to understand and as the father of nuclear physics, Lord Rutherford said, "These fundamental things have got to be simple." And, Werner Heisenberg said that, "Even for the physicist the description in plain language will be a criterion of the degree of understanding that has been reached."

The vortex theory for electric and magnetic fields is much simpler and far more understandable than the accepted version of quantum theory but it is not so useful to physicists. In the vortex theory I explain what space is but I don't provide supportive math so my contribution is not much use in helping scientists conquer space. Quantum theory doesn't tell us what space is, but as a system of mathematics it worked well in helping NASA fly rockets to the moon.

The vortex theory is a cross between physics and *metaphysics.* Metaphysics is concerned with what things are whereas physics is more concerned with how things work. Any success of the vortex theory would speak for the value of metaphysics in science; especially in science education.

In the vortex theory, vortex energy is considered to extend beyond our perception into infinity. If you cut a pie in half, then a quarter, then an eighth and so on you will not get rid of the pie, you will just get smaller portions of pie. So it is with the vortex. As the vortex extends out in three dimensions it gets thinner but never disappears entirely. This is why every vortex is treated as an infinite extension.

Vortices of energy are forms of activity so they are intrinsically dynamic. As they overlap they interact. In the vortex theory, electric and magnetic forces are treated as consequences of the dynamic nature of extending three dimensional vortices of energy.

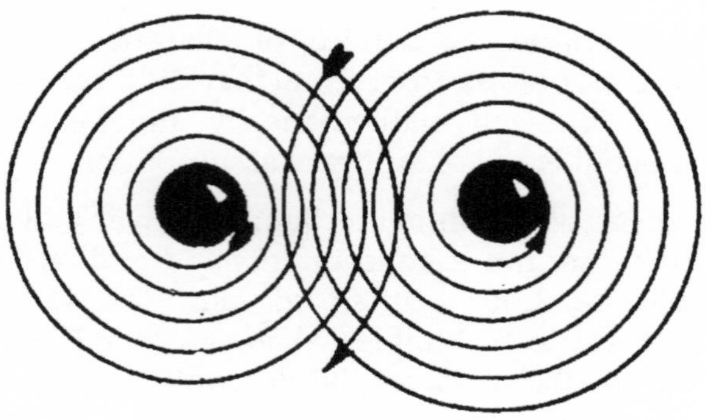

Space and atoms are electrically neutral because they consist of an equal number of particles that are opposite in electric charge. Opposite electric charges cancel each other out. In the vortex theory, opposite electric charge is set up by the opposite direction of spin either into or out of overlapping vortices of energy. This form of motion can be compared with a whirlpool.

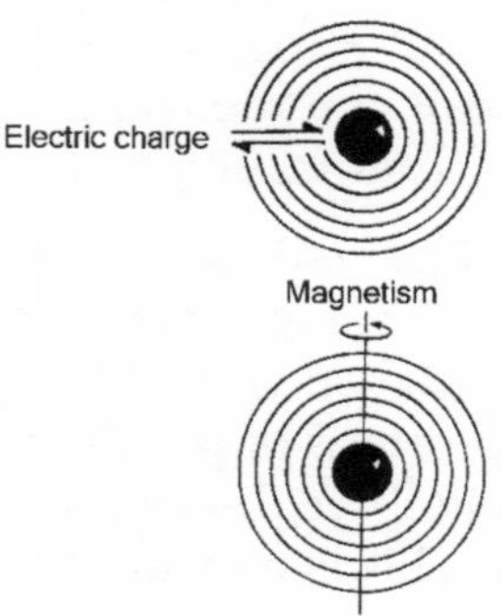

Magnetism is considered in the vortex theory to be caused by the rotation of vortex energy. This spin of the vortex can be compared with the spin of a roundabout or a top. The two forms of motion of the vortex, one superimposed on the other, are effective at right angles. The generation of electricity depends upon these force fields acting at right angles. Together the infinite extensions of vortex energy set up the two infinite extending fields of force associated with subatomic particles of matter; electric charge and magnetism.

In the classical field theory the forces between bodies of matter are thought to result from a universal field pervading space. In the vortex theory, the presumption of a single universal field is abandoned. Instead each vortex of energy, extending into infinity, is considered to be a *quantum field effect*. According to the vortex theory, the universal field is made up of individual *quantum field particles* of vortex energy that extend into infinity and interact with one another due to their innate dynamic state. The apparent universal fields of gravity, magnetism and electric charge, are considered to be the collectives of an infinite number of *quantum field effects*; that is, the addition of the effects of an infinite number of extensions of vortex energy, reaching out

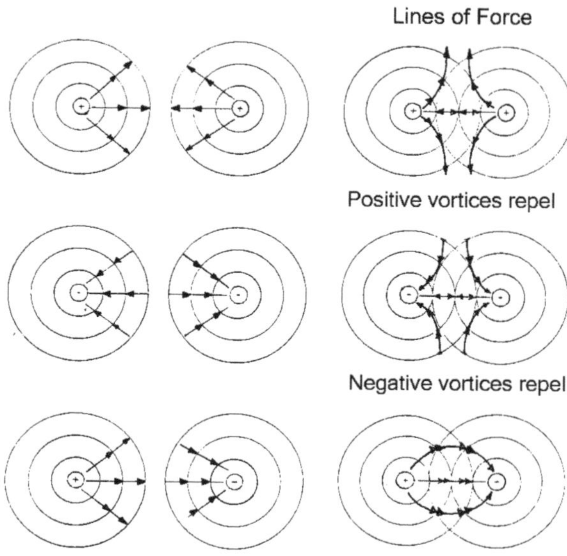

from the countless subatomic particles of vortex energy in existence.

Michael Faraday's *lines of force* can be explained in a simple diagram (on the previous page), which shows graphically how overlapping and interacting, extending concentric spheres of vortex energy, that are either growing or shrinking, could set up the classic *lines of force* associated with electrically charged particles. It also shows why it is that similar charges repel and opposite charges attract.

In the vortex theory light is not treated as electromagnetic radiation. The infinite extension of the vortex of energy is considered to be the source of the fields of electric charge and magnetism. These are properties of subatomic particles of matter, not properties of photons of light. The fact that photons of light interact with vortex particles of matter does not necessarily mean they are the same thing.

Democritus declared that *shape* was all important to the way atoms interact. If the *atom* of Democritus is perceived to be a particle of energy rather than a particle of material substance, and if the shape of the particle of energy is considered to be a vortex or a wave, then his insight could be used as the basis of a new approach to quantum mechanics.

CHAPTER 9
QUANTUM ROMANCE

Physicists have attempted to unify light and matter in their electromagnetic theory describing fields and waves. However, in the vortex theory, light and mass are treated as two distinct forms of energy which appear to display duality of a sexual nature. Indeed they appear to be quite romantic!

The wave from of energy is like sperm; it can be considered as the masculine form of energy, which is the basis of light and heat, X-rays and radio waves. The spherical vortex of energy is like the ovum; it can be considered as the feminine form of energy and the basis of subatomic particles.

The vortex theory reveals that the distinction of the sexes applies at every level of the Universe. That was one of the seven principles of Hermes Trismigestus. His first principle was that motion or the fluid state is the basis of all perceived reality. He also taught that patterns repeat at every level of the Universe; *as above so below, as below so above.*

Native Americans spoke of Grandfather Sun and Grandmother Earth. The word *matter* is derived from the Latin *mater* for mother and Lucifer, the light-bearer, was always imagined to be masculine.

The masculine *wave form* of energy is kinetic and can be likened to a traditional man who goes out to work. The feminine *spherical vortex form* is static and can be likened to a traditional woman who stays at home.

Just as the man penetrates the woman so the masculine, propagating wave form of energy penetrates the feminine static vortex form of energy. The *quantum,* driving into the spiral space path of an electron vortex, penetrates and then pushes it forward in wave motion and causes it to become a *wave-particle.* This is the vortex account for the *wave-particle duality* of subatomic particles.

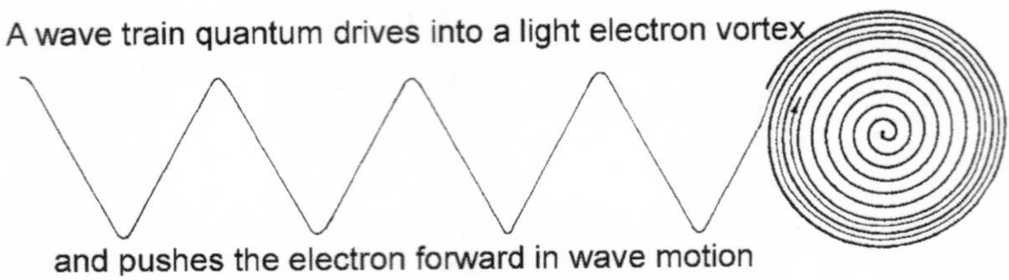

The wave particle duality of subatomic vortices can be appreciated by analogy with tadpoles. Just as the wavy tail of a tadpole drives the tadpole through the water in wave motion so the quantum sticking out of a vortex would drive the vortex forward in wave motion.

The proton vortex is massive so it can have sufficient inertia to withstand the impact of a quantum driving into it. Therefore it doesn't move forward and so a quantum can drive right into it and becomes *completely captured*.

The electron vortex is not very massive. It does not have sufficient static inertia to withstand the impact of a single quantum driving into it. The tip of the quantum gets tangled in the electron spiral space path but because the electron moves under the impact of the quantum it is only *partially captured*. Most of the quantum is left sticking out of it like a tail and because it cannot stop moving the *partially captured* quantum drives the electron forever forward in wave motion. This is why electrons behave as waves and it is difficult to know where they are from one moment to the next.

To understand partial capture, imagine the proton as a wooden gate post and the electron as washing on a line. Now imagine a man is asked by his wife to hammer nails into the post to fix the gate. He hammers a nail into the post. The post doesn't budge and the nail goes in a treat. But then he starts messing about and has a go at hammering a nail into her washing on the line. The hanging sheets keep moving under the impact of the nail, so the nail, apart from catching just enough to tear the linen, hardly drives into it at all. This is because the static inertia of the gate post is greater than the static inertia of the sheets on the line.

So it is with protons and electrons. A quantum can push an electron about but it will drive right into a proton. The interaction between wave-particles and vortex-particles is a tussle; it is a *tussle between opposing inertias.*

If subatomic particles of matter and quantum particles of light are treated as two distinct forms of the motion we call energy, then the endless interaction of these two forms of energy would be the constant tussle of opposites. The tussle is between *opposing types of inertia.*

As the kinetic inertia of the wave form of energy attempts to push the vortex form forward in wave motion, the static inertia of the ball vortex resists the attempt to be pushed forward. In the *tussle of opposing inertias* between the two forms of energy, the winner depends on which form of energy has the greater of its type of inertia. Everyone knows this to be true of relationships too. As in the quantum world, the dynamics of the tussles between opposing inertias also occurs between men and women in relationships.

A row ensued between the man and wife. As he couldn't get his way with her, he messed with her washing instead of fixing the gate. She then had a go at him because he tore her sheets with his nail. Shouting "I'll fix it," he rushed outside and instead of another nail he drove her car into the post.

Just as the wooden post would move under the impact of a car, while it wouldn't move under the impact of a nail, so a proton, being nearly two thousand times more massive than an electron, would require a quantum with kinetic inertia, more like that of a driven car than a hammered nail, to overcome its static inertia and push it forward.

Energy flows forever forward. It follows that a quantum wave train of energy driving into the spiral space of a vortex of energy could not back out. That is why the wave form of energy would get captured by the vortex form and why the marriage of wave and vortex – as in the ground state electron of an atomic orbit – is very stable.

The *partially captured energy* driving the capturing vortex forward accounts for its kinetic energy. But the captured energy

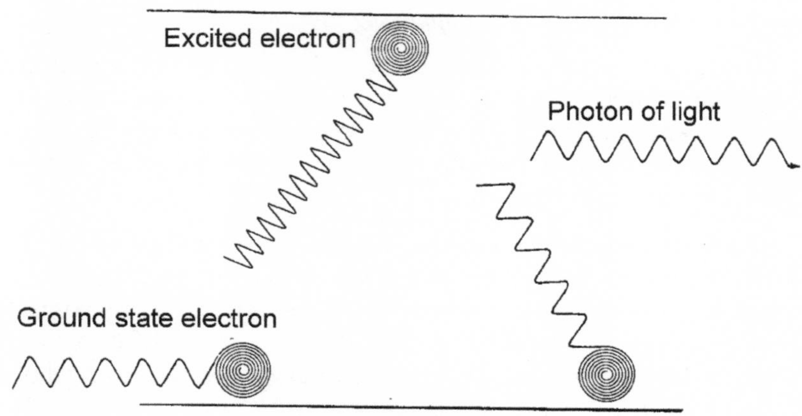

would also add to its mass. That explains why subatomic particles become more massive when they are accelerated. The *partial capture* of energy can also explain the quantum leap.

The *ground state electron* is an electron in its normal orbit in an atom. If the electron absorbs energy it can leap into a higher orbit – also known as a higher quantum state - in the atom. It is then called an *excited electron.* The excited state of an electron is unstable so the excited electron may fall back to the ground state again and release the energy it absorbed as a photon of light – with a frequency equal to the difference in energy between the two orbits. This is what happens when you strike a match. The wood ignites and burns. The heat released by the combustion excites electrons in carbon atoms, floating in the hot air currents, rising from the burn. The excited electrons leap to a higher orbit then drop back to the ground state and release energy as photons of light, which you see as the flame.

Imagine a fellow who has a romance with a married woman. When they are first together it is very exciting but the romance doesn't last. She goes back to her husband and is grounded and her lover leaves her to become an old flame. That happens in carbon atoms whenever we strike a match and see a flame.

The fellow's wife divorced him because he cheated on her and she started a romance with another man. She didn't ditch the new man however. The chemistry was right so she married him. This happens in the quantum world too. It is the basis of chemistry.

When, in a chemistry lab, a test tube is placed in the flame of a Bunsen burner, the lady electrons, in the atoms of the mix in the test tube, are picked up by quantum lovers, they then get excited and take a quantum leap. A chemical reaction occurs because after the quantum leap, instead of ditching their quantum boyfriends, they go off with them to live in a new atomic home. Just like the ex-wife who went off to live in a new home with her new husband.

The fellow and his ex got on okay after their divorce so there was no problem in her visiting her old home. In chemistry the link formed between atoms, where electrons nip back and forth between the old atomic home and the new one is called a *covalent bond*. Covalent bonded atoms form *molecules*.

Some people, after a divorce, don't get on so well. They don't pop in an out of each other's homes; instead she moves into the new home and never returns to the old one and acrimony is a charge between the old and the new homes.

In chemistry the link of electric charge between atoms, after an electron leaves one atom for another and never returns, is called an *ionic bond*. Ionic bonded atoms form *salts*.

Men without women can be destructive. They get together in gangs and go off to war wrecking everything and everyone in their path. In physics that is called *entropy*.

Women without men become frigid. That is the form of entropy where matter without energy to heat it up freezes.

When men marry women, not only do they keep each other warm in bed, but the man's energy gets channeled into work. Swords are fashioned into plough shears and men go to work instead of to war. This happens in physics too. When matter is heated, useful things happen. We get to enjoy hot showers and baths, tractors till the land and delicious food is cooked on the stove. In physics this opposite to entropy is called *work*.

Richard Feynman said of quantum mechanics, "You never understand it, you just get used to it."[1] Quantum mechanics is difficult to understand when quantum reality is treated exclusively as waves of energy. As soon as we accept the vortex as a second, distinct form of energy alongside the wave form, then the mechanics of quantum reality becomes easier to understand. It is important to appreciate however, how the neutron fits into this approach to quantum theory.

1 Calder N. *Key to the Universe: A Report on the New Physics* BBC Publications 1977

CHAPTER 10 UNDERSTANDING THE NEUTRON

A question remains as to why neutrons are unstable. If electrons are attracted to protons by their opposite charge, why do they fall apart so easily when they are outside of an atomic nucleus?

According to *The Vortex Theory*[1] the answer to this conundrum lies in the fact that the electron in a neutron has energy in *partial capture* equivalent, according to $E=mc^2$, to one and a half times of its rest mass. The *kinetic inertia* of this energy would appear to be sufficient to enable an electron to overcome the *static inertia* of its electrostatic attraction to the proton and break free from entrapment in the neutron.

In the nucleus of an atom, if an electron in a neutron breaks away from the proton it could be caught by another proton. The result of this would be that all the protons in the nucleus, as well as the protons in the neutrons within the nucleus, would be acting to keep the electrons in the nucleus. Outside the nucleus, without the support of other protons, singleton neutrons would be more prone to lose their electrons.

In *The Vortex Theory* I used an analogy of cats and mice to illustrate this point. I likened electrons in the nucleus to mice and the protons to cats. Most atomic nuclei have more proton cats than electron mice so as soon as a mouse breaks away from a cat it is caught by another cat which doesn't have a mouse. As a result of this predominance of cats the mice would be unlikely to break free of the group. However, a cat on its own with a very lively mouse would be more likely to lose the mouse than if it

1 Ash D. *The Vortex Theory*, Kima Global Publishers, 2015

were in a group of cats continually pouncing on the mice as fast as any escape.

Now if you think about it a mouse, or a cat with a mouse in its jaws, would be more likely to break out of a larger, looser group of cats and mice than a smaller, tighter one. This is because the larger groups would be less coherent and with more space between the cats there would be more chance for a determined mouse to escape. The loose group might even lose cats as well as mice. That was how I used this analogy to explain why atoms like uranium, with a large, loose nucleus, tend to be radioactive. Their radioactivity is caused by their losing *Beta decay electrons* – mice in the analogy – and *alpha-particles* – cats and mice in the analogy.

Single escaping mice pay a price however. As a single mouse breaks free of a group of nuclear cats – in *Beta decay* – they lose a bit of tail to tooth or claw. That was how I used my cat and mouse analogy to account for the nebulous particles called *neutrinos.* Neutrinos were proposed to account for energy loses that occurred in the decay (*Beta decay*) of neutrons in radioactive atomic nuclei.

The nucleus of a rare atom of uranium (Isotope U^{235}) has a metaphorical cat-flap that lets in any stray cat, with a mouse in its teeth (a neutron) that is looking for a home. However, when the neutron stray enters with its mouse, it disturbs the equilibrium of the house. The feline congregation first becomes turbulent, then it swells and finally it splits in two, releasing a lot of pent up energy from the group (nuclear energy). At the same time two or three cats, each with a mouse in its jaws, get kicked out. These strays then look for new homes and if they find more of the rare uranium atoms, with the allegorical cat-flaps, they race in and disrupt them releasing more felines, fire and spitting fury.

Physicists in the Manhattan Project isolated, purified and concentrated the rare atoms of $uranium^{235}$ with the allegorical neutron cat-flaps and then crushed them into a ball. As they did so a chain reaction began as two or three cats with mice in their jaws (neutrons) split two U^{235} nuclei to release four or six more jawful cats which split around four U^{235} nuclei to release eight neutrons then sixteen, thirty two and so forth, rapidly disrupting ever more nuclear households. With each split two or three felines in a flurry were released until the neutron strays caused millions more nuclear homes to split up then in an exponential flash, the number grew to countless trillions.

In the chain reaction, over Hiroshima, it took only thirteen thousandth of a second, for neutrons to penetrate and split the nuclei of practically all the $uranium^{235}$ atoms in the first atomic bomb.

BOOK III

THE VORTEX COSMOLOGY

If we do discover a complete theory, it should in time be understandable in broad principle by everyone, not just a few scientists. Then we should all, philosophers, scientists, and just ordinary people, be able to take part in the discussion of the question of why it is that we and the universe exist. If we find the answer to that, it would be the ultimate triumph of human reason – for then we would know the mind of God.

Stephen Hawking

CHAPTER 1
DISCOVERING THE VORTEX

When I stumbled across the idea that the vortex of energy was the basis of matter, I never realised I had also discovered a new way of appreciating space-time, a breakthrough in understanding gravity, the origin of antimatter and the source of dark energy. No one knew about dark energy in 1965 when I first read about the vortex of energy in the *Advanced Course in Yogic Philosophy.*[1] When I borrowed the book from an attic antiquarian library in 1965 I didn't know the author was an American, William Atkinson, (1862-1932) writing under the pen-name Yogi Ramacharaka. I imagined an Indian ascetic in a loin cloth scribing in a cave.

It was not the authorship of the book that impressed me, however, it was the publication date. I was amazed to see the book had been published in 1904, a year before Albert Einstein unleashed his famous equation $E=mc^2$ on the world. I may have been only sixteen but I knew enough about physics from my dad and my reading – I wasn't taught physics at school – to appreciate the importance of someone anticipating Albert Einstein in publishing that matter is a form of energy.

It struck me that if the author of the book knew, before Einstein, that matter is formed of energy then his description of how energy forms mass might be correct. He said that according to Yogic insight, energy; *prana* exists in matter in the form of vortices; *vritta*. That was the premise that started me on a life long adventure with the *vortex of energy*.

1 Ramacharaka Yogi, *An Advanced Course in Yogi Philosophy*, Fowler 1904, (Cosimo Classics Philosophy, 2005)

During my first term at Queen's University in Belfast, in 1968, Professor Gareth Owen, Dean of the Faculty of Science, said to me that he was disappointed that students treated university as an extension of school, a place to cram for exams in order to get high grades and a good job. He impressed on me that university was a place for personal development. Universities provided a platform for young people to go beyond the rote learning of school to think original thoughts and develop ideas and new theories.

I was galvanised. I abandoned my plans for a career. I determined to develop a new theory instead. I knew I had been prompted to work on the vortex so I started to read to see how the vortex theory might sit with modern physics.

Lord Kelvin – who was born in Belfast – had fielded a theory of the vortex atom. It failed because in his day the atom was thought to be the smallest particle of matter. I knew the smallest particles of matter were not atoms; they were subatomic particles. I decided to see if the vortex hypothesis would apply to electrons, protons and neutrons. I would investigate to see if these subatomic particles were vortices of energy.

I realised, almost immediately, that the opposite charges of protons and electrons could be accounted for in terms of opposite direction of spin in or out of the vortex. A vortex of energy could not exist without charge because electric charge was an expression of its intrinsic, dynamic state.

The third particle in the atom, the neutron, didn't have a charge. In my new vortex theory it had to be an electron bound to a proton. Its neutral state must have come from the opposite directions of spin, of its constituent electron and proton cancelling out each other's dynamic effect. The neutron had to be an electron bound to a proton because I was taught in physics that neutrons formed when electrons combined with protons, they had the sum mass of an electron and proton and that they decayed into an electron and a proton.

I was only a month into my project when one of my physics teachers told me that neutrons were definitely not electrons bound to protons. He didn't say why. He also disappointed me by saying that subatomic particles could not be vortices because a vortex has an axis of spin whereas subatomic particles appeared to be spheres without axial poles. Bang went the theory.

Crestfallen I abandoned my work on the vortex but I couldn't stop thinking about it. Then, back in Cornwall during the Xmas break I had a breakthrough. I had been billeted in a caravan with a man called Harry who was staying with us at the time. One evening Harry asked me to hold up a shank of wool for him while he wound it into a ball. Harry was a bachelor and he knitted his own socks.

As I watched Harry wind the wool I had an epiphany. As the ball of wool grew in his hands I saw a vortex forming as a sphere with no *measurable* poles because the axis of spin was changing all the time. On my return to Queens I went back to work on the vortex theory with renewed enthusiasm and within weeks I had developed an account for antimatter.

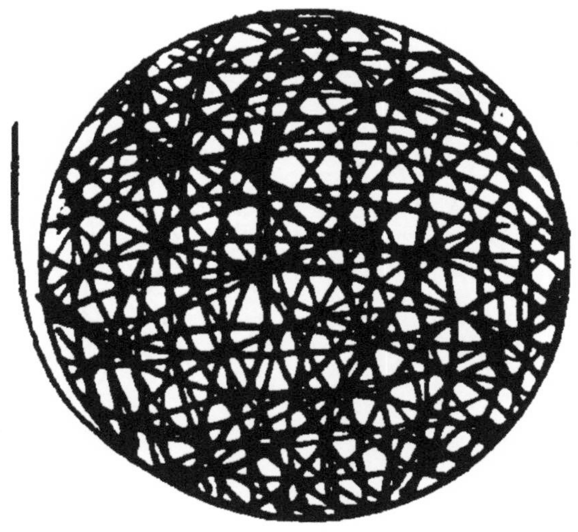

CHAPTER 2
ANTIMATTER

During my final year of A-levels at Hastings College of Further Education, prior to going to university, I developed a model for the quantum of energy – also known as a *photon* of light, heat or gamma-rays.

I treated light, not as vibration in a ubiquitous electro-magnetic field but as two lines or strings of vibration propagating at right angles to each other to form a quantum or a photon. I was taught that the energy in each photon of light was proportional to the number of waves it contained. Blue light was more energetic than red as it had more waves in each of its photons. I constructed a model of the quantum or photon as two undulating lines of the movement of light, each one propagating at right angles to the other.

I drew a diagram for the quantum as two perpendicular wave trains of energy. I was nineteen at the time. When I was twenty, and at university, I used my diagram of the quantum to illustrate how antimatter was formed.

In my reading I discovered that when atoms of lead were bombarded with gamma rays, some of these exceedingly high energy quanta passed through the nuclei of the atoms. When they did so one wave train of energy in the gamma ray quantum transformed into an electron and the other wave train transformed into a *positron* – a positively charged electron; a subatomic particle of antimatter.

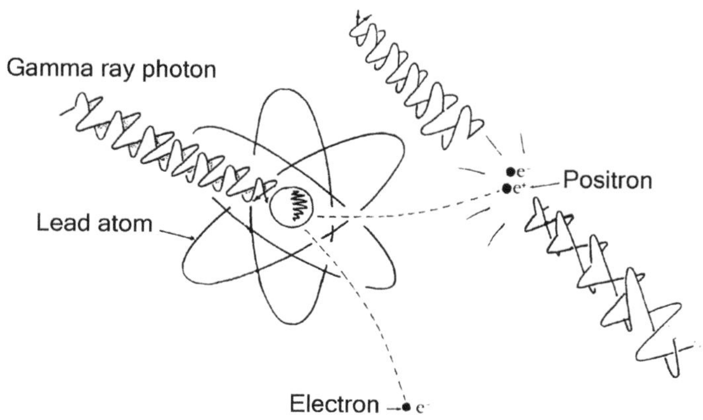

I drew a diagram to illustrate the two lines of energy in a gamma ray quantum being transformed into two new vortex particles. This confirmed my view that energy existed in two interchangeable forms; vortices and waves. The diagram also supported my hypothesis that when energy was forced through a vortex it could be transformed into a vortex. This occurred as the lines of the movement of light, in a gamma ray quantum, passing through vortices in the nucleus of an atom were forced to spin. Each line of energy was transformed from the wave form into the vortex form. The two wave trains in the gamma ray quantum were thereby transformed into two vortices.

My vortex hypothesis enabled me to explain, in very simple terms, how energy could be transformed into mass. But the antimatter experiment showed not only how mass could be created out of energy; it also revealed how mass could be destroyed and transformed back into the wave form again. The newly formed positron was attracted to electrons by virtue of opposite electric charge but as soon as the positron accelerated into an electron they both annihilated. (Only mass was annihilated. Energy was not annihilated)

I accounted for the *annihilation of matter and antimatter* as two vortices of energy, with the same mass but opposite direction of spin (opposite sign of charge), 'unzipping one another'. As they unwound each other from the vortex to the wave form the two lines of energy, that had formed the positron and electron vortex, reverted to the radiant form of energy. (After annihilation two quanta of gamma radiation move away from the site of annihilation in opposite direction, each with half the energy of the original quantum.)

One line of the movement of light had formed an electron of matter whereas the other had formed a positron of antimatter. That suggested half the quantum of energy is presumptive matter and the other half is presumptive antimatter. Later I discovered that in fluorescence and polarisation only one of the fields in the quantum reacts with matter and in photography only one field of energy in light reacts with the photographic plate. It seemed that half of every quantum does not belong to the world of matter at all.

CHAPTER 3
THE QUANTUM COIN

I imagine energy as lines or strings of the movement of light. It is not literally a line or string. The line is an imaginary model I used to explain the form and action of energy.

If a line were folded into waves, in a bundle of fixed size, the more waves in the bundle the greater would be the length of line and therefore the more energy it would contain. The quantum, as a bundle of waves of light, is governed by a constant called *Planck's constant* – named after the originator of quantum theory. In physics this is represented by the symbol 'h'.

At nineteen, when I drew a diagram to show that the more waves in a bundle of fixed size the more energy it would contain, I realised the *constant of proportionality* for a 'bundle of energy' containing only a single 'imagined' line of the movement of light would have to be '½h', half Planck's constant, because a quantum containing two wave trains is governed by 'h', the whole of Planck's constant.

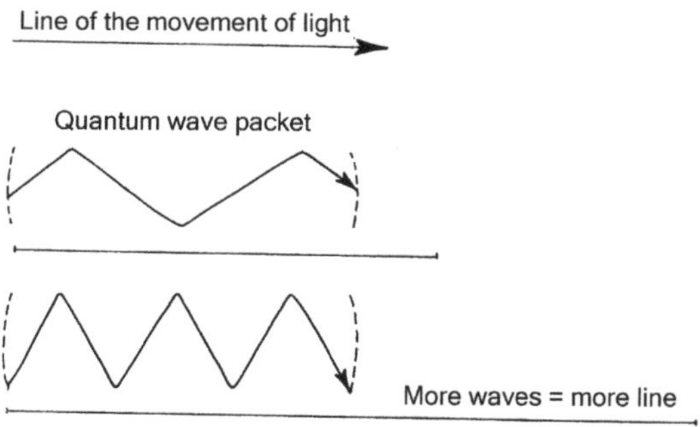

Quantum events are governed by half Planck's constant i.e. ½h or h/2. It was much later I realised this was because only half the quantum is effective in the world of matter. It was fifteen years before I was able to explain why quantum events occur in integer multiples of half Planck's constant.

I now imagine the *quantum coin*, as the basic currency of the Universe, to be a single 'imaginary' line of the movement of light. The denomination of the quantum coin is, ½h.

I visualise each vortex of energy to be formed of a single 'imagined' line of the movement of light spinning. That is how I explain why electrons and protons are governed by ½h; half Planck's constant.

A quantum of energy is formed out of two 'imagined' lines of movement of light, (as is a photon, which is a quantum of visible light). That is why, in my model, the quantum and the photon are governed by the whole of Planck's constant.

I depicted energy as money. To begin with in physics the quantum was thought to be the basic currency in the Universe; the energy penny, so to speak. But then physicists discovered, when dealing with the energy pennies, that only half a quantum penny had value in this world. The other half penny in each quantum seemed like a foreign currency. It appeared to belong to another world. Physicists called the quantum penny a *Boson* named after the Indian physicist Satyendra Nath Bose. They called the quantum half-penny a *Fermion* after the Italian physicist Enrico Fermi.

CHAPTER 4
THE MIRROR UNIVERSE

I have no problem in understanding the conundrum of the quantum coin because in my vortex physics the Universe has to be arranged in two identical, mirror symmetrical halves, one as matter and the other, antimatter. The quantum coin conundrum suggests to me that half the quantum is in our matter half of the Universe and the other is in the antimatter half. The split of energy between two halves of the Universe leads me to believe that the two halves would have to be *mirror symmetrical, not only in form but also in action.*

I began to develop the idea of a parallel Universe of antimatter in the summer of 1969 when I was twenty. It happened as I watched a stream that runs through the valley of Crackington Haven in Cornwall where I lived with my family (A documentary *Westward TV* programme, *Summer of the Ash Family,* now on You Tube shows this scene in Cornwall. It was filmed for the *ITV* network in 1972).

As I was thinking about the flow of energy in the vortex I was wondering where the energy was coming from and where it went to as it spun in or out of the subatomic particles of matter. As a student of physics I abided by the law that energy is neither created or destroyed so I could not accept the idea that energy could just pop up out of nothing and then disappear into nowhere.

While sitting on the bank of the stream, watching the water tumble by, an idea popped into my mind. It struck me that the energy, passing through the vortices and waves that made me and everything I could see, was somewhat like the water in the stream. The water kept moving through the whirlpools and ripples in the stream because it was part of an endless cycle of water circulating between ocean, rain, spring and stream then back to the ocean again.

As I stared at the stream at the bottom of our garden it was apparent to me that the waves and vortices in it were not solid and real. They were just forms that were maintained by the constant flow of water. It dawned on me, as I watched the turbulence in the water, that everything in the world was just an illusory form maintained by the constant flow of energy.

The question that dominated my mind was where did the energy come from and where did it go to. As I pondered on the stream I realised that the energy forming everything and everyone in our world of matter, could only flow continually if, like the water in the stream, it was part of an endless cycle. To my mind energy like water had to circulate.

Just as the water cycle started with water molecules rising from the sea, so I envisaged vortex energy emerging from the centres of subatomic particles. Just as the water became first cloud, then rain which filled the stream before returning to the ocean again, so I imagined energy, forming the world in which we live, to be in a Universal cycle of vortex energy.

I began to develop my model for the Universal cycle of vortex energy by drawing a diagram of energy spinning into the centre of a subatomic vortex as though it were a funnel, passing through the centre as though it were a tunnel and then spinning out to form a *Siamese twin vortex,* with equal mass but opposite charge, on the other side of singularity.

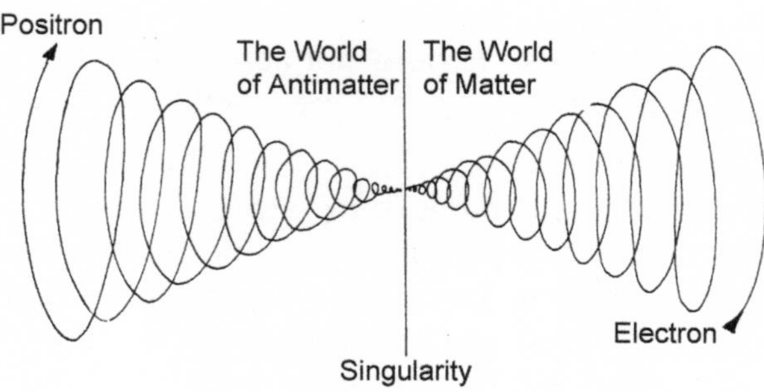

For energy to form a uniform cycle it would have to circulate through every vortex of matter into an equal but opposite vortex of antimatter. That would mean beyond the centre of every subatomic particle of matter there would be a particle of antimatter. This implied that every form and action in our world of matter would be faithfully replicated in a mirror symmetrical world of antimatter. The parallel world of antimatter had to exist beyond the centre of everything for my vortex hypothesis to work.

If this line of thinking is correct then there would be an antimatter version of you, in an identical mirror half of the Universe, reading an antimatter version of this book. The question to ponder is which is the real you and which is the mere reflection?

If energy circulates between vortices of matter and antimatter then in each quantum there must be a cycle of energy. It was decades before I realised that energy might be circulating in the quantum between the two undulating lines of the movement of light.

That idea only came to me when I realised one wave train of energy, in every quantum, was presumptive antimatter which took no part in the world of matter we occupy. That was also when I realised why quantum events incur only half a quantum of energy rather than a whole quantum. I now imagine the half quantum that is not involved in our world could be the reverse cycle of energy.

The pattern had already emerged in my mind in 1969 that each identical, mirror symmetrical, half of the Universe acts as a reverse cycle of energy for the other. This arrangement allowed for the endless cycle of energy. It also enabled me to predict a Universe which is finite, because it does not stretch on endlessly, but also is infinite because it has no boundaries.

A dual state of being finite and also infinite can be understood from the Earth. We know the Earth has a finite size but if we were to keep journeying on it, we would never reach the ends of the Earth, we would simply arrive back where we started from.

Just as people in the Northern hemisphere of the Earth are completely surrounded by the Southern hemisphere, and vice versa – they could never leave one hemisphere without entering the other – so each half of the Universe is surrounded by the other half. It would be impossible to leave the Universe of matter without entering the Universe of antimatter and vice versa. But each identical sphere would be separate from the other and both would have a finite size. My model of finite and yet infinite Universe fits with the *3-sphere model* proposed by Albert Einstein in 1917.

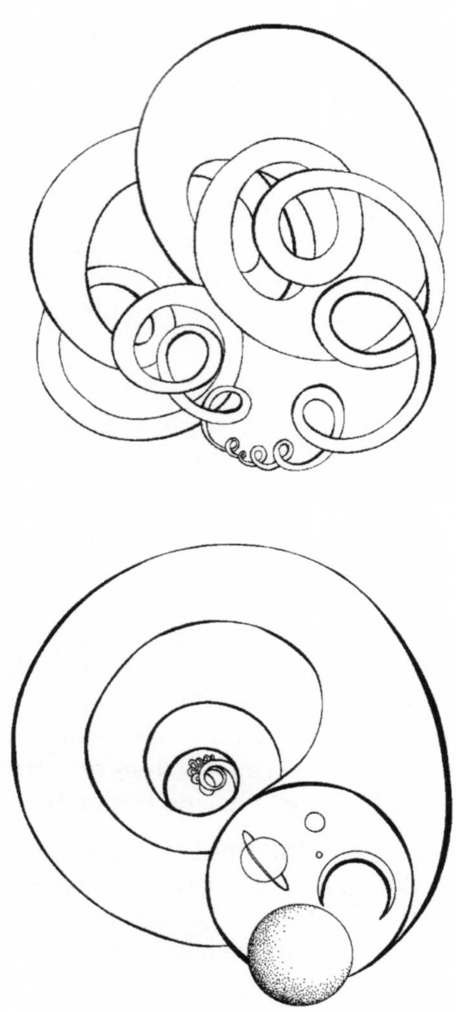

CHAPTER 5
SPACE

In 1969, the troubles broke out in Northern Ireland so I transferred from Queens University in Belfast to Queen Elizabeth College, London University. It was there in 1970 where I began to develop the *Vortex Cosmology* based on the idea that space is an infinite extension of vortex energy.

It made sense to me that space is the infinite extension of vortex energy because space extends in three dimensions and in my hypothesis the vortex of energy is responsible for all three dimensional extensions in the world we live in. As matter and space are both three dimensional extensions I presumed that space must be connected to matter in some way. An unusual incident, in 1970, triggered in my mind the realisation that matter is vortex energy we perceive and space is vortex energy extending beyond our perception.

In 1974 I later discovered I was on the right track when I read a new book, *Einstein: His Life and Times.*[1] According to the book, when Albert Einstein went to New York in 1919 a reporter asked him, as he stepped ashore, to put his theory of relativity in a single sentence. His reply was, "If you remove matter from the Universe you also remove space and time."

Einstein's response to that reporter could only mean one thing; matter, space and time must be connected. Matter, space and time may not be different things they may be different expressions of the same thing. I realised the difference between matter and space may not be a difference in what they are but a difference in the way we perceive them.

If my thinking was correct then space, like matter, had to be particulate. Back in the early seventies I enthused with the other

1 Clerk R.W. *Einstein: His Life and Times,* Hodder & Stoughton, 1973.

students about a *quantum theory for space*. I used my *ball of wool* model to explain the quantum theory for space. I explained that the spherical vortex of energy could be imagined as a ball of wool stretching into infinity. Just as a ball of wool appears as a spiral at the centre and then more and more like a sphere as it extends out from the centre, so mass would represent the dense spiral centre of the vortex of energy; whereas space and force fields would be more like concentric spheres of energy stretching out into the distance from it.

Decades later I discovered that idea of space and force fields, as extensions of the spherical vortices of energy, was depicted in the ancient Egyptian symbol: The Flower of Life.

The Flower of Life diagram depicts extending concentric spheres of vortex energy – each a quantum continuum of mass, space, charge, magnetism and gravity – overlapping each other. That is why *The Flower of Life* is rightfully considered to be the diagram that universally represents everything.

My idea was that the vortex energy extending from every subatomic particle of matter in a body would set up *concentric shells of space* surrounding the body. I also spoke of *bubbles of space*. The bubble of space extending from a body I imagined as an extension of the shape of a body somewhat like an *aura*.

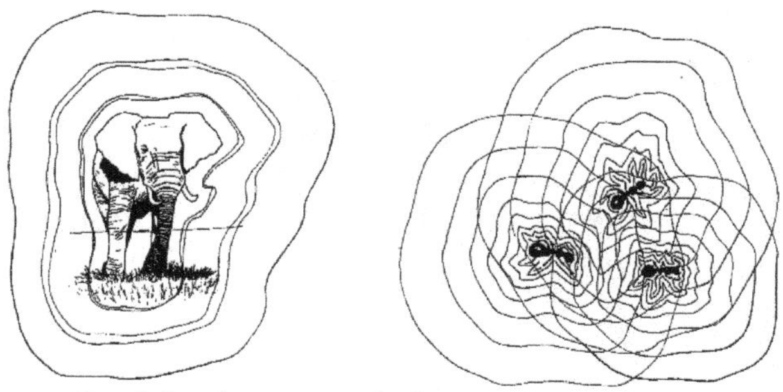

Space is the vortex energy extending from bodies of matter into infinity

The idea I had was that everyone and everything in the Universe is surrounded by their own bubble of space, which moves with them wherever they go.

I used this idea to explain Einstein's premise in his special theory of relativity that *the measured speed of light is independent of the velocity of the observer*. It was obvious to me why that would be so. If we measure the speed of light, the photons whose speed we are measuring would be in our own bubble of space which moves with us as we move.

Measuring the speed of light *relative to our own space* would cause the measure to be independent of our own movement because *our own space would be moving with us*.

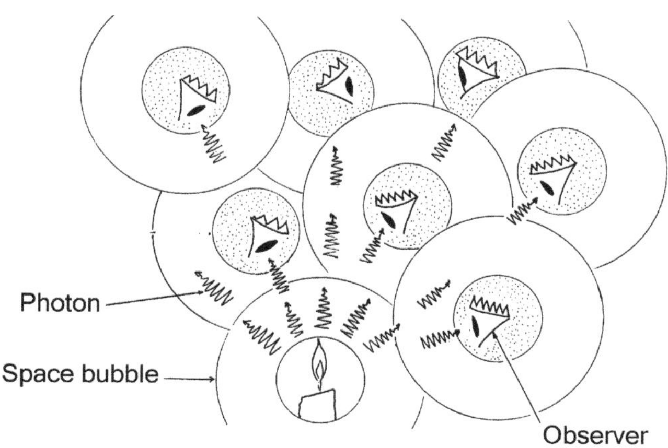

When Albert Michelson and Edward Morley carried out their classic experiment in Cleveland Ohio in 1887 to see what effect the movement of the Earth would have on their measure of the velocity of light, they were measuring the velocity of light *relative to the Earth's bubble of space,* which, as an extension of the Earth, was moving with the Earth. It is hardly surprising that they discovered their measure of the velocity of light was independent of the velocity of the Earth.

I got my understanding of this result, which has been described as the Second Scientific Revolution, back in 1970, from a dachshund called Johannes. Yes you did read it right. I got my breakthrough understanding of one of the greatest developments in the history of science from a dog. No the dog didn't talk but the dog did have piles and they led me to my epiphany.

In my first two years as an undergraduate student at London University I had digs in Kensington square near Holland Park. I often took Johannes, the family dachshund, out for a walk around the square. On the day of momentous revelation, unbeknown to me, Johannes was ailing with piles. Poor Johannes was wrapped up in a blanket. The blanket roll, was intended to confine him as a 'sausage dog' to his wee doggy bed in the front hall.

As I burst through the front door a spectacle beheld me that was to lead to one of the most important breakthroughs in my life. Johannes, still wrapped in his blanket, was off his bed and was pulling himself across the floor toward me with his front paws. He obviously wanted walkies but he couldn't pull himself out of his blanket roll because his blanket was moving with him. (Johannes did recover from his piles.)

Mrs. Weir, his mistress, came running up to retrieve the little fellow but not before I had my mental thunderclap. Johannes represented a moving body. The blanket roll represented the space extending from the body. The body couldn't move *relative* to its own space because its space went with it as it moved. In that moment I got relativity.

As I went on developing my theory of space, I thought about the *real* concentric spheres of vortex energy flowing through *apparent* concentric spheres set up by the levels of intensity of energy. These levels of energy intensity were dictated by distance from the centre of the body of vortex energy from which the space originated.

While the intensity of energy of each growing or shrinking concentric sphere of vortex energy was constantly changing, the levels of energy through which they moved never changed. They were determined purely by distance from the centre of the vortex.

I realised the *fixed levels of intensity of energy* that set up the manifest forms of matter and space were responsible for what the Yogis described as *maya*; the illusion of forms. Every day the table is there for your breakfast because its form is stable due to this effect. Like a *mirage* these fixed levels of intensity of vortex energy are responsible for the illusion of material substance in the physical world.

I explained this point to Mrs. Weir by analogy with the large Catholic family to which I belonged. When I was little, for a number of years, more or less every year a baby had been born into the family. The babies grew into children and as each one of us grew out of an age there was always another brother or sister growing behind us to take our place so that each 'age year' was always occupied by someone.

The babies represented energy, like bubbles, popping out of the singularity point at the centre of the 'maternal vortex'. The child growing through each year of age corresponded to a real bubble of energy growing in size through each level of intensity defined by its distance from the centre of the vortex. These levels of intensity of energy set up the *forms* of matter and the *shells* of space, which persisted, not because they were real but because there was always a sphere of vortex energy at each level of intensity.

I got this idea from watching the flow of water through the ripples in the stream in Cornwall. The water in the stream was real but the ripples were not. They were forms that seemed real but remained only while the water flowed through them. I saw this as the flow of the *real* water in the stream through the *apparent* forms of the ripples.

In each ripple it was never the same water from one moment to the next. So in the vortex of energy from moment to moment it is never the same concentric sphere of energy at each level of intensity. This was illustrated in the family analogy. Every year there was a ten year old in the family but from year to year it was a different child that occupied that position. As fast as the ten year old grew to eleven there was a nine year old growing to ten. A dotty old granddad might think he was meeting the same ten year old at Christmas but if he did he would be deluded.

We are like the dotty old granddad. We may think that matter is solid and substantial, after all the things in our homes are there all the time so they seem real enough. However, their stable reality is an illusion because the vortex energy flowing through them is not the same energy from one moment to the next.

The concentric spheres of vortex energy flowing through the levels of intensity are real. However, it is the levels of intensity of energy that set up the *forms* of matter and the concentric *shells* of space. These levels are not real things they are just mathematical parameters. The seemingly static appearance of space and the substantial appearance of matter are illusions set up by the fact that energy is continually flowing through the levels of intensity that set up matter and space. This is the illusion of materialism and we are deluded by it to imagine the unreal as real.

The concentric spheres of vortex energy setting up the space extending from each body of matter would extend into infinity. As they do so they would drop in intensity. However, this diminishing in intensity would be compensated for by the addition of vortex energy from the space extending from other bodies. In that way Universal space could be the addition of all the many bubbles of space extending from every vortex of energy in existence

I imagined the vortex energy extending from the Earth, and then combining with the vortex energy that extends from the Sun and all the stars and galaxies in the Universe, to set up the vastness of space. I understood the *universal cycle of vortex energy* by analogy with the water cycle. The largest sphere of universal space was to me the equivalent of the ocean. The smallest spheres of space, the equivalent of springs, represented the singularity points at the centre of each subatomic vortex of energy.

In the reverse cycle of energy, through the antimatter half of the Universe, there would be shells of space extending from the antimatter Earth and antimatter Sun that would combine with the space extending from all the other antimatter stars and galaxies in the antimatter half of the Universe. That would set up a vast shell of space, which would be the outermost and the largest sphere of space for antimatter.

Just as the Atlantic Ocean is connected to the Indian Ocean so I imagined the largest shell of space extending from matter would be connected to the largest shell of space extending from antimatter. As two halves of a whole, the universe of matter and the universe of anti-matter would face each other over the same vast frontier of space. Between the maximum and minimum spheres of space, vortex energy would flow between the matter and antimatter halves of the Universe.

In the water cycle water circulates between the large and the small, between raindrops and oceans. So it would be in the cycle of vortex energy. Vortex energy would circulate, through matter and antimatter, between the extremes of smallness and largeness; between the smallest and largest sized spheres of space.

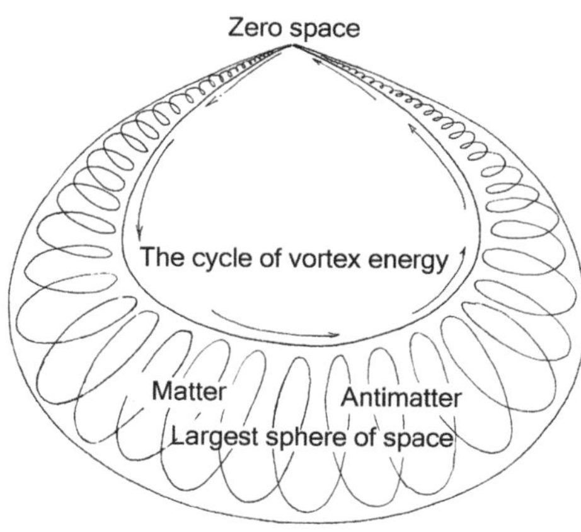

One of the main things that came from Einstein is that there is no absolute space. I realised space is made up of countless quantum bits or bubbles of space. Each quantum of space is an extension of a particle of vortex energy into infinity. I used this principle to explain to my father how his dowsing worked when he used a pendulum to dowse for water underground. I explained to him the moving space, extending from water flowing or rising underground, would deflect his pendulum from its swing through the otherwise non-moving space, extending from the ground under which there was no water flowing or rising.

In the vortex theory the photon of light is considered to be a totally different form of energy to the vortex. It has no mass and it is not subject to extension in three dimensions. The photon moves in the space set up by the vortex of energy but it does not form space. As the energy circulates in the photon, forwards in matter and backwards in antimatter, it exists in both halves of the Universe simultaneously. This could help to explain why photons are not constrained by the extension of space but are seemingly everywhere all at once.

One day, as I looked out of the window of the college library at the traffic on Campden Hill Road, I realised that each vortex and wave train of energy, as a system of motion, exists *relative* to every other vortex of energy as an extension of space. Motion could not exist unless there is space, just as the cars couldn't drive up Campden Hill without the road. So each and every vortex particle of energy depends on every other vortex extension of energy for its existence. I called this co-dependence between particles of energy: *The Universal*

Law of *Love,* i.e. everything in the Universe depends on everything else for its existence.

Each vortex particle of energy acts as an individual system of motion *relative* to adjacent particles of energy acting as bubbles of space. The wave packets of light then move through the *space foam* extending from countless trillions of vortices of energy. Light depends on the three dimensional extension of vortex energy for its existence as particles of wave motion.

I imagined *space-foam* somewhat like frog spawn. The massive centres of the vortex particles were like the black frog embryos at the centre of each egg, and the space extending from them was like the frogspawn foam. Each frog egg was part of the collective spawn of all the other eggs. Each egg had its own integral three dimension of space forming a sphere. As I thought about the baby frogs growing in their 'eggs of space' I realised they were growing in a *fourth dimension of space.* I began to wonder if maybe there might be a fourth dimension to space in which everything grows. The tadpoles and everything that grows could be growing in a fourth dimension of space, from little to big.

The idea of a fourth dimension of space came to me in Kensington Gardens. I was already charged with excitement at the realisation that there were images of my vortex cosmology in *Alice's Adventures in Wonderland.* The identical twins *Tweedle Dum* and *Tweedle Dee* agreeing to do battle fitted with particles of matter and antimatter annihilating each other. The story of Alice chasing time – in the watch of the white rabbit – down the rabbit hole and then shrinking in size in order to enter a looking glass world, configured with my idea of shrinking space leading into a mirror symmetrical world of antimatter. Her journey, after drinking the *drink-me-drink,* was in the fourth dimension of space.

I envisaged the fourth dimension, associated with space, to be the dimension of bigness and smallness in which energy accelerated, into the vortex as one sign of charge or out of the vortex as the opposite sign of charge. I called the fourth dimension of bigness and smallness the *Alician Dimension.* I named it after Alice in wonderland.

In the three dimensions of space things move up or down, forwards or backwards, left or right. In the fourth dimension of space things move in all three dimensions simultaneously as they shrink or grow. The vortex of energy grows or shrinks – according to its sign of charge – in three dimensions simultaneously. That to me was the fourth dimension of contraction and expansion of vortex energy. It was the dimension in which things could shrink or grow.

Backwards and forwards, left and right, up and down are the ways we move through the three dimensions of space. We move through the fourth dimension of size of space, the *Alician Dimension* when we grow from a single cell into an adult human body.

The vast distances of space are an illusion set up by the Alician dimension of size. Intergalactic space is only vast to us because we are little compared to a galaxy. It would take forever for a bacterium to cross the road because it is tiny compared to us whereas we can nip across in a moment. Likewise if a galaxy had legs, in galactic time it would not take long for it to stroll across to a neighbouring galaxy.

Time is linked to size. It is also linked to acceleration in the vortex of energy, which stretches in the fourth dimension between the smallest of the small and the largest of the large.

In the vortex cosmology I visualised every vortex of energy extending in three dimensions to form space. The fourth dimension of size of space was set up by the growing or shrinking of concentric spheres of energy.

Every vortex exists as a system of motion *relative* to the extensions provided by other subatomic vortices of energy. That *relationship* of vortex motion, in the fourth dimension of size of space, relative to the extending three dimensions of space, provided by other vortices of energy was what I perceived as time. Because shrinking size of space and acceleration are features of the vortex, it follows that time would be relative to size and to acceleration.

The infinity sign depicts the two halves of the Universe and the infinite cycle of energy between them.

CHAPTER 6
TIME

The time we measure is set by mankind to help us define our relationship with the world in which we live. Time enables us to synchronize events and is a necessary construct in complex societies.

To help us measure time we observe the movement of the Earth and the moon in space. A revolution of the Earth on its axis is our day and its period of traverse around the sun is our year. The days we divide into twenty four segments, which we call hours and hours are divided by sixty to get our minutes and sixty again to get our seconds. So long as we all agree to stick by these rules, time works for us all. But this time isn't real. As Yuval Noah Harari said in his book *Sapiens: A Brief History of Humankind*:[1]

"Any large scale human cooperation…is rooted in common myths that exist only in people's collective imaginations…"

Time would appear to fit Harari's category for a myth and as such time is arbitrary. For example, the Catholic Church changed the rules for months set by the moon. Faced with awkwardness because the phases of the moon in a year fitted the Witches favourite number thirteen – which is why it was the Witches favourite number – Pope Gregory rigged the calendar to twelve months. Extra days were added to eleven months and leap years were used to top up the twelfth with an extra day, every four years.

As I worked on the vortex theory for space I realised there is no absolute space and there is no absolute time. Each vortex of energy, as motion, relates to every other vortex as space and *quantum time* defines this relationship.

1 Harari Y. Sapiens: *A Brief History of Humankind,* Vintage, 2014

I realised time is not a thing. Time is the relationship between things. My interest was in *quantum time* that defined the relationship between vortices of energy, and vortices and waves of energy. I realised the fourth dimension of size of space would impact *quantum time* because the vortex of energy was, in essence, a system of altering size.

Albert Einstein proclaimed time to be a fourth dimension in continuum with the three dimensions of space. I accepted Einstein's idea only when applied to a *quantum space-time continuum* because the vortex is, in essence, a quantum continuum of space-time.

As motion in the vortex flows simultaneously through all three dimensions of space, time would be associated with this relationship between movement and space in the dimension of size. That is how I understood time to correspond to the fourth dimension. In the fourth dimension of shrinking and growing, time would also shrink or grow.

I reasoned *quantum time* would be influenced by the *Alician dimension* of size of space at every level in the Universe. The subatomic world represented one zone of size in the *Alician dimension*. Outer space was just another zone.

I imagined a giant in the zone of outer space to whom the Earth could be a football. I visualised him using the Earth as his clock. I envisioned him taking the rotation of the Earth as his seconds and the circuit of the Earth round the Sun as his count of hours. Our clocks would race for such a space-giant. For a much bigger 'creator of the Universe', looking down on her entire creation as a football; two billion years to man would be but one of her days.

I accounted for the 'seven days of creation' in terms of the dimension of size. If time were relative to the size of space, billions of years to man would be but a few days to a 'Creator of the Universe'. Indian philosophers compared a phase of Universal expansion not to a week but to an inspiration of breath.

Remember when you were little how long days lasted and how weeks stretched out, especially when you were waiting for a birthday? Then when you grew up time began to race. Weeks seemed shorter and months would fly by and there was no longer enough time in the day to pack everything in like you did when you were a kid.

The periods of time set by mankind apply to the size-zone average adults occupy in the fourth dimension. I predicted the time periods set by man would *appear* to differ according to size. According to my premise, *relative* to the periods of time set by human convention, time would appear to dilate should we shrink into a smaller zone of space and contract should we grow into a larger space zone in the *Alician dimension*.

I rarely watched television in my youth and almost never when I was at university but on one occasion, as I strolled through the student union, I was mesmerized by a TV. It was showing a programme on small animals and insects that had been filmed and then the films had been slowed down. Flies appeared on the wing like birds, and cats moved with the grace of panthers. As I watched the programme I realised that the movements of flies and cats might appear to me to be jerky and fast because I was bigger than them but in their zone of space-size they would appear to each other to be as birds and panthers are to us, in our zone of size.

It struck me that quantum time would be relative to the size of quantum space. If the size of a single quantum of space was set by the vortex – with extension from the centre of the vortex the concentric shells of space are progressively larger – so time would be relative to each zone of size set by the size of these concentric shells of space.

Using the simple equation that speed is equal to distance divided by time I knew that time had to be related to space, represented by distance, and the speed to the speed of light. I began to suspect that time must be related in some way to the flow of *real* vortex energy through the *apparent* levels of intensity of vortex energy; the shells of space.

I appreciated that the movement of energy in the vortex from one level of intensity to the next was a period. But the periods were not fixed. The periods would be longer, with expansion from the centre of the vortex, as the concentric spheres of energy grew larger. The periods would be shorter with contraction into the centre of the vortex of energy as the concentric spheres shrank smaller. In the vortex of energy, the motion into the centre is acceleration.

My work in 1972, on time and space, coincided with my brothers and I signing a contract with a music publisher to record our songs. When we were in the studios recording some of the tracks, I was transfixed by the spools of tape fast winding on the recording decks. As the tape wound off a spool at speed, the spool accelerated. I saw dramatically, how spin in a vortex could cause acceleration.

Everything was beginning to make sense for me. The motion of energy into the vortex is acceleration. That explained to me why time was effected equally by acceleration as by size.

As I watched the spools of tape accelerate in the recording studios I realised acceleration in the *Alician dimension,* into the centre of a vortex of energy, could have the same effect on a large or a small observer's experience of time, relative to the human measures of time, as acceleration in a rocket ship toward the speed of light.

This was clear to me when I read Einstein's twin paradox theorem. According to Einstein if one of a pair of twins were to accelerate in a rocket ship, almost to the speed of light, the time they live through would be dilated compared to the time lived by their twin left on Earth. The time dilation would slow down their rate of growth. The result would be that on reunion, after the journey, the space twin would still be a small child whereas the twin who was left behind would be a full grown adult. This bizarre effect has been confirmed in particle accelerators. *Muon particles* accelerated close to the speed of light lasted ten times longer than muons at rest.[2]

There is no universal time. If we could accelerate to the speed of light, our time would dilate into eternity. If we could accelerate our energy beyond the speed of light we would

2 Calder N. *Key to the Universe: A Report on the New Physics,* BBC Publications, 1977

transcend physical time altogether and experience multi-dimensional reality. Even in the constraints of physicality, in meditation or heightened awareness, we can transcend time and meld into the eternal now. We can experience eternity now if we go beyond our perceived time.

Time is not universal. Time is localized. Every vortex of energy, as a system of motion, exists relative to every other vortex as a quantum of space and quantum of time. Each vortex would, therefore, be a quantum space-time continuum relative to every other vortex particle of energy.

I also realised time is not just the relationship between vortices of energy. Time is also set up by the relationship between waves of energy and vortices of energy. Energy flows through space as light and heat. That flow sets up time. The direction of light, always going forward and never going back could set up the flow of time we experience as the past, through the present and into the future.

If the forward propagation of energy through space sets up forward time then the reverse flow of energy in the quantum would set up reverse time relative to us. If the reverse flow of energy in the quantum is effective in the world of antimatter then time should flow in reverse relative to us in antimatter. In fact it does. Time does appear, from the human standpoint, to be running in reverse in antimatter. However, this is not a cause for concern. We see words reversed in mirrors but that does not mean they are actually reversed. It is just an effect of mirrors. The flow of time, which appears to us to be reversed, is actually flowing forward in the 'mirror-symmetrical' world of antimatter.

Another factor that effects time is gravity. The time recorded by a very accurate clock, in the realm of a crawling baby closer to the centre of the Earth, is dilated relative to the time measures of adults. Time experienced by little people is marginally different to that of big people. The time dilation associated with smaller size, stronger gravity and the twin paradox have something in common: acceleration.

The fourth dimension of *size of space* is the dimension of acceleration into the vortex, or deceleration out of the vortex.

Gravity is a force of acceleration. Gravity is a centralizing force acting in the fourth dimension of space that causes the acceleration of bodies into smaller space. Gravity may also affect the passage of time because of acceleration. However, I do not believe that gravity is caused by the warping of space-time by mass, as Albert Einstein supposed. In the vortex theory gravity would appear to be caused by something completely different.

CHAPTER 7
GRAVITY

I was not happy with Albert Einstein's theory that gravity results from the distortion of a universal, four dimensional continuum of space-time by mass. After his quantum initiative in 1905, I was disappointed at his return – in the *General Theory of Relativity* of 1915 – to the classical concept of a ubiquitous, universal field of space and time. Despite the sensational success of his theory of general relativity it set up difficulties in integrating gravity into the quantum theory that he himself had established.

A way of relating gravity to the quantum continuum of space-time had to be found. I was confident I could find it in the vortex theory. Amazingly, a new theory for gravity emerged quite quickly from the vortex theory that seemed to coalesce with quantum thinking and could also explain the curvature of space that Einstein had predicted in his theory.

I reasoned that the quantum of energy exists either as a 'packet of waves' or a 'whirlpool of spin'. The *quantum of vortex energy* is 'mass' at its centre and 'space' as it extends into infinity. Electric charge, magnetism and gravity are expressions of vortex energy at a distance. They are all causes of acceleration because acceleration is a feature of the vortex.

Electric charge is made up of unitary charges contributed by individual particles of matter. Gravity is the same. It is made up of the contributions of gravity from each and every particle of matter. Gravity, like charge, is a three dimensional extension - operating in the fourth dimension of size.

I reasoned that gravity had to be a feature of the vortex of energy because if its association with mass and space and because acceleration, caused by gravity, is a hall mark of the vortex.

It was obvious to me that gravity arose from the quantum world as a quantum effect. It appeared to behave as a collective of quantum effects. The gravity in every body of matter seemed to be a collective of the components of gravity contributed by every subatomic particle in a body of matter, according to its mass. And gravity appeared to be acting from the centre of mass of each and every particle of matter. I couldn't see why Einstein didn't base his theory for gravity on quantum theory.

Einstein was my icon. I was impressed by his 1905 pioneering contribution to quantum theory and his special theory of relativity, which treated every system of energy as being *relative* to every other system, rather than to arbitrarily, assumed parameters. His original thinking was very much quantum thinking and laid the foundations for quantum theory; especially in terms of the *relational* interactions between quanta appreciated by Heisenberg. His crowning glory was $E=mc^2$ where he predicted that mass is a form of energy. All I did, with the vortex model, was to show how energy forms mass. I was then able to show *how* vortex energy could cause gravity.

I didn't like Albert Einstein's arbitrary idea that gravity is caused by the distortion of a space-time continuum by mass. He predicted that a distortion of space-time around the sun, caused by its mass, would result in an apparent deflection of starlight around the sun. Einstein suggested the effect could be observed during a solar eclipse.

The Royal Society launched two expeditions in 1919 to test Einstein's prediction. They took photographs of stars around the sun during a solar eclipse in two locations. They then photographed the same stars in the night sky when the sun

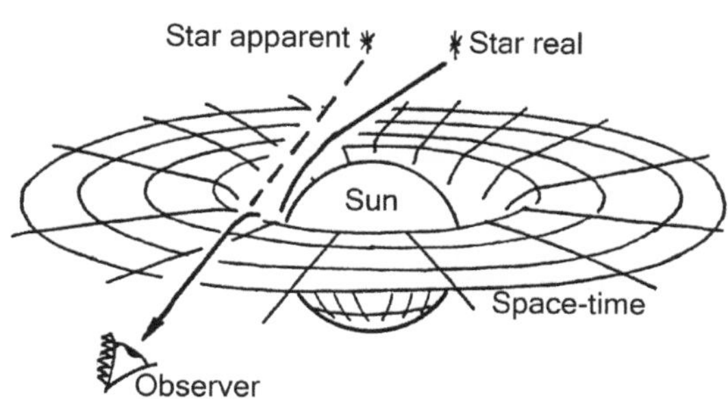

wasn't there. When they compared the photographs they saw the positions of the stars close to the sun had changed precisely as predicted by Einstein. Obviously it was assumed the stars had not changed their positions. It must have been deflections of starlight during the eclipse, caused by the sun, as predicted by Albert Einstein, which caused the effect. Overnight Einstein became the most celebrated scientist in the world.

In his general theory of relativity Einstein predicted that space-time around the sun was *distorted* by its mass which he declared to be the cause of gravity. In the vortex theory I was able to show the same effect of the sun on starlight if space is perceived to be an extension of mass.

According to the vortex theory, space in the immediate vicinity of the Sun would be an extension of the spherical shape of the Sun. The Sun would be surrounded by vast concentric shells of space. These would cause the space around the sun to be *actually* curved.

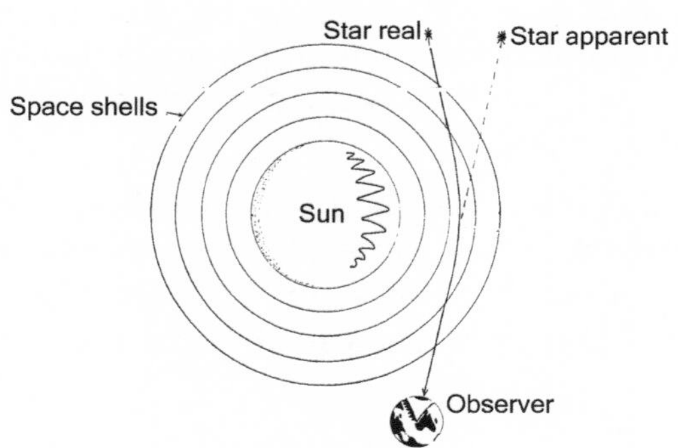

My interpretation of the 1919 Royal Society observation of the deflection of starlight around the sun, during a solar eclipse, was that space is not an empty void but is a form of energy extending from a body of matter. The shape of space is an extension of the shape of the body it extends from. In the case of the sun that would be concentric spheres.

The concentric spheres of space, extending from the sun, which I described as *shells of space,* would cause the actual curvature of space around the sun. Light from distant stars, following this curvature of space, would be *deflected* round the sun because the space near the sun acts like a roundabout.

Photons travelling from a distant star could be likened to cars on a road. When they came to the sun in their path, they would encounter the equivalent of a roundabout in space. The photons would then follow the curvature of space around the sun as surely as cars on the road would follow the curvature of the road around a roundabout.

You can test the *deflecting effect* of space extending from a body of matter, for yourself, by swinging a pendulum along the edge of a table, a couple of inches from it. The pendulum swing will follow the line of the table. When you come to a corner of the table the pendulum will change direction of swing around the corner of the table and then continue to swing in parallel to the next edge of the table beyond the corner. The swinging pendulum, following the contours of space extending from the table will track the outline of the table *at a distance from it*. If you perform this experiment around other objects you should see the same effect.

By offering an alternative account for Einstein's prediction of the curvature of space around the sun I was in a position to challenge his *general theory of relativity* that gravity is caused by mass distorting a supposed universal space-time continuum. I was then positioned to propose a new theory for gravity, based on the vortex theory.

When I was in the recording sessions watching fast wind tape accelerating the spools, on reel-to-reel recording decks, I saw in a graphic way how acceleration toward the centre of a spiral is a feature of the vortex. Gravity causes things to accelerate toward the centre of the Earth. The recording sessions coincided with my developing an account for antimatter. I began to think that *'maybe there is a link between gravity and antimatter'*.

If gravity rises from the centre of the vortex of energy, maybe it is an effect arising from antimatter. My vortex theory only made sense if there was a vortex particle of antimatter with equal mass but opposite charge beyond the centre of every vortex particle of matter. If my vortex theory was correct I knew that there had to be antimatter in equal amounts to matter in the Universe. That fitted with Lord Rutherford's *Law of Conservation of Electric Charge* which suggests that there are an equal number of opposite charges in the Universe.

I knew that particles of matter and antimatter are opposite in charge and there is an *accelerating force of attraction* between opposite charges. So I began to think that just as electric charge was a vortex interaction, acting beyond the surface of subatomic particles, maybe gravity was a vortex interaction acting from beyond their centres.

I knew that matter and antimatter experience a force of attraction, caused by electric charge. Maybe it was that same force, acting through the centres of bodies of matter and antimatter, which causes them to accelerate toward each other, in toward the smallest realms of space. This centralizing force would draw them in towards a point of singularity where they would meet. And, I reckoned, where they meet, if the gravity is strong enough, that was where matter and antimatter could annihilate and generate *annihilation energy*.

I realised that the pull between matter and antimatter would not be between the *connected twin particles* because they are a single system of energy. These connected vortex particles could not interact relative to one another. The force of attraction, leading to annihilation, would have to be between a body in matter and a body in antimatter where they are not connected and therefore they can move relative to each other.

For example there would not be a vortex interaction between my body in matter and my body in antimatter because they always move in parallel therefore they could not move separately. However, the mass in my body of matter could experience an accelerating force of attraction, through the centre of the matter Earth, to the mass of the antimatter Earth. At the same time my twin body of antimatter could be experiencing an accelerating force of attraction, through the centre of the antimatter Earth, toward the matter Earth.

My account for the gravity of the Earth that keeps us all attached to it is not that the Earth is distorting space-time, rather each one of us is being pulled toward the centre of the Earth. And that pull is coming from the antimatter Earth through and beyond the centre of our own planet.

Einstein invented a story for gravity. I invented a different story. His story and mine both work to account for the deflecting effect that massive bodies, like the sun, have on starlight travelling close by them. Both stories account for the accelerating force of gravity. The question is which story are we to believe?

We are not bound by what other people believe; even if that is the consensus view. As my father said, "Even if you find yourself standing alone against the whole world it doesn't mean you are wrong."

My story for gravity requires that you believe in the existence of an antimatter Earth and that you are pulled by a force of attraction to the antimatter Earth through the centre of the Earth. I cannot prove my version of gravity. It simply follows a line of emerging logic in my theory. However, I believe in it because it has enabled me to explain 'unexplained mysteries' in outer space and in the Earth.

When matter meets antimatter, annihilation can occur between them. As I followed this line of thought I realised gravity could be tapping into the most powerful source of energy in the Universe.

The energy release of nuclear fusion in a hydrogen bomb or the nuclear furnace of the sun comes from the burn of 0.7% of the mass of a proton. In annihilation of matter and antimatter, 100% of the mass of a proton is consumed to release nearly two hundred times as much energy per proton as involved in the process of nuclear fusion.

I reasoned that because vortices of energy exist in each other's space, they could all, ultimately, coincide on the space of a single proton. Theoretically the entire Universe could collapse, due to gravity, onto a singularity point less than the size of a pin head. However the force of repulsion between protons would work against gravity to prevent that from happening. Protons packed in the nucleus of an atom do not undergo annihilation because the force of gravity pulling them together is weaker than the force of electric charge pushing them apart.

For the annihilation of protons to occur naturally the matter and antimatter would have to be sufficiently dense for the force of gravitational attraction to overcome the force of charge repulsion. The only situation I could think of where this might occur would be in a black hole.

In the early 1970's when I was working on the idea of annihilation energy being released by gravity in black holes, I thought that maybe *quasars* could be evidence of this effect.

Quasars had recently been discovered and were then considered to be the most energetic things in the known Universe. Releasing around two hundred times the energy of stars, the energy profile fitted matter-antimatter annihilation.

Quasars had been discovered at the furthermost reaches of the Universe and at the centres of galaxies. Both of these places where were I expected the matter and antimatter halves of the Universe to meet. I had ascertained that black holes could transform into quasars. Once matter-antimatter annihilation ignited they could change from the darkest pits to the brightest lights in the Universe.

In January 1975 I presented my vortex theory and vortex cosmology at the Royal Institution and then in March I got married and settled down to build a house and earn a living. I kept working on my theory and my wife drew charming drawings for my first book, *The Tower of Truth* [1] for children. But I lost enthusiasm for the vortex theory because I had not yet found a way to explain the strong nuclear force. That came a decade later.

Even when I picked up the vortex theory again, I let my cosmology lie fallow as I focused on nuclear physics and my challenge to quark theory. Then, when I was working on my first book with Peter Hewitt, which came out in 1990,[2] Peter drew my attention to the potential 'neutron undermine' of the uncertainty

1 Ash D & A, *The Tower of Truth*, Camspress, 1977
2 Ash D & Hewitt P, *The Vortex: Key to Future Science*, Gateway Books, 1990

principle. The focus on challenging quantum theory took my mind further from the vortex cosmology.

I had parted from my first wife and married again. With my second wife I embarked on an international lecture tour which would last until 1994. Only then, when writing *The New Science of the Spirit*,[3] did I revisit the vortex cosmology. That book, published in 1995, included the vortex account for quasars, the challenges against quantum mechanics, the quark theory, the big bang and general relativity theories and the prediction, in a chapter on the *Vortex Cosmology*, of an acceleration in consequence of a vortex action, of the further-most galaxies away from us, the effect now known as *dark energy*.

My second marriage ended in 1996. In 1997 I married again and went into business with my third wife. That marriage ended in the summer of 2006 and in the autumn of that year I discovered cosmology had caught up with me. Reading *Just Six Numbers* by Martin Rees,[4] I realised my prediction of the *dark energy effect* had been vindicated by a team observing super nova explosions in distant galaxies.

The dark energy discovery, of the acceleration away from us of the most distant galaxies, was published in 1997, two years after I published my prediction of this effect. Realising that my vortex hypothesis was now a sound scientific theory, I decided to pick up my 'brain child' and tend to it again.

By the time I published *The New Physics of Consciousness*,[5] astronomers had discovered that black holes at the centres of galaxies were erupting periodic bursts of gamma rays.

My excitement grew as I knew gamma rays were the signature of matter-antimatter annihilation. Not only did the

3 Ash D, *The New Science of the Spirit*, The College of Psychic Studies, 1995

4 Rees M. *Just Six Numbers*, Weidenfield & Nicolson, 1999

5 Ash D. *The New Physics of Consciousness*, Kima Global, 2007

gamma ray bursts offer solid proof of matter-antimatter annihilation, occurring through the most dense and powerful gravitational centres, but this discovery led me to predict in *The Vortex Theory,*[6] that in the emission of gamma rays, black holes were behaving somewhat like *Old Faithful* in Yellowstone Park; black holes were releasing gamma rays in the manner of a geyser.

The cause of periodic gamma ray bursts from galaxy cores was obvious to me. Gamma rays would be produced continuously, by matter-antimatter annihilation, through the centre of a black hole, at the centre of a galaxy, as matter and antimatter are compressed by gravity into singularity. But captured in the black hole, acting like a gargantuan vortex, the gamma ray photons would not be immediately free to fly. Instead they would build up in concentration, as captured energy, until the amount of captured energy exceeded the ability of the black hole gravity to hold it. Then the energy would escape as a burst of gamma radiation and the cycle of capture would begin all over again in the action of a geyser.

A black hole can only act as a *gravity geyser of gamma radiation* if it had a continuous feed of gamma rays. That suggests an antimatter black hole exists beyond singularity in the core centre of the black hole. These gamma ray bursts led me to conclude not only that the vortex cosmology was correct, but that we are heading toward an earth convulsing conclusion!

In 2012 I met Susan Saillard-Thompson. She had a copy of Immanuel Velikovsky's *Worlds in Collision,*[7] a book I read in my student days which I had never forgotten. Velikovsky, a Russian geo-scientist, had scoured the records of practically every ancient culture and discovered a common theme. According to ancient records the sun has repeatedly changed its direction in the sky and the changes are invariably associated with global

6 Ash D. *The Vortex Theory*, Kima Global Publishing, 2015

7 Velikovsky I. *Worlds in Collision,* Victor Gollancz, 1950

catastrophes that wipe out entire civilisations and plunge tiny remnants of survivors back into a Stone Age.

Then more recently I acquired a copy of Charles Hapgood's *Earth's Shifting Crust.*[8] Hapgood suggested that repeated massive cataclysmic changes afflict our planet due to periodic shifts in the Earth's crust. In the foreword to this book, Albert Einstein said he was electrified by Hapgood's work and verified it was a sound science that accounted for periodic pole shifts, sudden and dramatic climate changes, mountain formations, the rising and sinking of continents and other unexplained mysteries in the Earth's history.

In his foreword Einstein raised the question as to where the heat in the Earth came from that would enable the crust to move freely over the inner layers of the Earth. Hapgood gave reference[9] to the heat in the Earth being one of the greatest unsolved mysteries in science.

Not only does the heat in the Earth require explanation, the periodicity of that heat rising from the core of the Earth, capable of causing repeated massive Earth changes, also has to be accounted for.

I was not convinced by Velikovsky's argument that the entire Earth rolls head over heels from time to time. The Earth is way too massive for that. Neither am I impressed by the idea amongst geo-physicits that the magnetic poles of the Earth can suddenly and unaccountably reverse. The massive iron core of the Earth is too massive to flip. Speculation by geo-physicists that the heat in the Earth comes from the spontaneous decay of naturally occurring uranium is also unsatisfactory, in my opinion.

8 Hapgood, C., *Earth's Shifting Crust: A Key to Some Basic Problems of Earth Science,* Pantheon Books, 1958

9 Gutenberg B., *Internal Constitution of the Earth,* Dover, 1951

Stephen Hawking's prediction, in 1971,[10] that black holes could occur in bodies with mass less than the mass of a star, led me to wonder if there might be a black hole at the gravitational centre of the Earth. If that was so then perhaps there could be a *gravity geyser* in the core centre of the Earth. That could explain, not only the heat in the Earth but also its periodic release from the centre of the planet.

I predict that annihilation between matter and antimatter is occurring through the singularity point in a *sub-stellar black hole* at the core centre of the Earth. Gamma rays resulting from annihilation are trapped, initially, by the gravity of the black hole but with the continual input of this annihilation energy, the black hole eventually becomes super-saturated with gamma rays. At that point they burst out of it.

The gamma rays would not escape from the dense, massive iron core of the Earth, but would be absorbed and in the absorption process their energy would be transformed into heat. Heat of the iron core would weaken the Earth's magnetic field. At the same time, heat would rise up from the iron core and into the mantle of the Earth under the crust.

When the rising heat reaches the layers of magma under the Earth's crust it would reduce the viscosity of the molten rock changing it effectively from glue that holds the crust in position to a lubricant that would enable it to move. Think of this effect somewhat like warming honey that turns it from 'sticky honey' to 'runny honey'. The rising heat under the crust would also cause 'global warming'. By the time the heat reaches the crust, the Earth's core could be cooling again.

The axis of rotation of planet Earth happens to be titled by 23.5° so at every solstice one pole is closer to the sun than the other. This list would set up a gravitational differential so that at the solstice the sun would be pulling stronger on one pole than another.

10 Hawking S. *Gravitational collapsed objects of very low mass,* Monthly Notices of the Royal Astronomical Society, 1971

The continual deposition of ice on the poles would increase the mass differential on the poles, which would increase the pull of the sun on the pole facing it during a summer solstice. Year by year, during a period of global warming when the crust is effectively floating on the inner layers of the Earth, the likelihood of the crust rolling 'head over heels', to reverse the poles, would increase. An 180° roll of the Earth's crust over the inner layers would cause not only a magnetic pole reversal but an apparent reversal of the direction of the sun in the sky. With the deposition of ice greater at a pole under land than sea, this cataclysmic event would more likely occur at an Antarctic summer solstice.

CHAPTER 8
DARK ENERGY

The attraction between matter and antimatter, across the largest sphere of space, could cause the largest things – galaxies – to accelerate into larger space. This decentralizing acceleration of matter, caused by vortex energy, could account for what is now known as *dark energy*.

My first presentation of the vortex theory, at the Royal Institution of Great Britain in London, on January 15th 1975, was to a monthly gathering of young members. I had over a hundred guests, so mine was the first ever meeting for Young Members of the R.I. in the main lecture theatre. I presented the vortex cosmology and – predicting the phenomenon now known as dark energy – I said the furthermost galaxies from us would be accelerating away from us faster than the ones that are closer to us because they would be nearer the largest sphere of space, therefore they would be experiencing a stronger pull from antimatter than the closer galaxies.

I explained that galaxies accelerate away from us because of a force of attraction between matter and antimatter over the largest sphere of space. I said this was a similar vortex interaction to that which caused the force of electric charge causing opposite electric charges to accelerate together.

I published this prediction, now commonly known as *the accelerating expansion of the Universe*, twenty years later, in 1995. That was in the chapter on the vortex cosmology in *The New Science of the Spirit*.[1]

In December 1997, Saul Perlmutter, professor of Physics at Berkley University, published the results of observing supernova explosions in distant galaxies. Prior to then it was

1 Ash D. *The New Science of the Spirit,* The College of Psychic Studies, 1995

assumed – from the big bang theory – that the Universe is expanding at a uniform rate. Perlmutter's observations[2] established that more distant galaxies are accelerating away from us faster than closer ones, which suggests the expansion of the Universe is accelerating. Perlmutter coined the term *dark energy* to indicate, "We are in the dark as to what causes it".

In *A Brief History of Time,*[3] Stephen Hawking said:

"A theory is a good theory if it satisfies two requirements: It must accurately describe a large class of observations on the basis of a model that contains only a few arbitrary elements and it must make definite predictions about the results of future observations."

Hawking's first criterion for a good theory was first proposed by a philosopher, William of Ockham in the 14th century as: 'non sunt multiplicanda entia praeter neccessitatem', i.e. "Entities are not to be multiplied beyond necessity." William of Ockham argued that we should accept as true the theory that has the least arbitrary assumptions. He argued this law of economy of ideas with such sharpness that it came to be known as *Ockham's razor.*

The vortex theory satisfies Ockham's razor by explaining the entire body of physics from the simple assumption that the Universe is formed of two distinct, inter-changeable and interactable 'particle types' of energy based on vortex and wave motion. From this single axiom I propose that:

- Mass is quantity of vortex energy.
- Static inertia originates from the spin of energy in the form of a spherical vortex.
- Kinetic inertia originates from the propagation of energy in the form of a 'packet' train of waves.

2 Perlmutter S. *Discovery of a Supernova Explosions* Lawrence Berkeley National Laboratory, Dec 16, 1997

3 Hawking S. *A Brief History of Time,* Bantam Press, 1988

- Potential energy is accounted for as vortex energy.
- Three dimensional extensions of matter, forces and space arise from the 3D extension of vortex energy.
- The forces of electric charge, magnetism and gravity are caused by the interactions of vortices of energy.
- Infinite extension of these forces and space result from the infinite extension of the vortex of energy.
- Electric charge is caused by expanding or contracting concentric spheres of vortex energy.
- Magnetism is caused by the rotation of expanding or contracting concentric spheres of vortex energy.
- Kinetics originates from the interactions between wave packet quanta of energy and vortices of energy.
- Wave particle duality is a consequence of a bound state between vortex and wave particles of energy.
- Space is the infinite extension of the vortex of energy beyond our direct perception.
- Space curvature is caused by the collective of vortex energy extending into infinity from all the vortex particles of matter in a spherical heavenly body such as the sun.
- Universal Space is the collective of vortex energy extending into infinity from all existing vortex particles.
- Electrons are light vortices where the concentric spheres of vortex energy are contracting.
- Protons are massive vortices where the concentric spheres of vortex energy are expanding.
- Neutrons are a bound state of electrons and protons.
- Nuclear vortices of energy have the capacity to capture quantum particles of propagating energy by virtue of being a spiral space path for them to travel into.
- Mesons are formed out of energy captured in proton and neutron vortices of energy.

- Nuclear binding is caused by captured energy swirling between vortex particles in atomic nuclei.
- Nuclear energy is the release of excess captured energy as nuclear vortices converge.
- Sort lived particles in high energy experiments are synthesized unstable, transitional, vortices of energy formed by the forced passage of energy through natural (proton) vortices of energy in atomic nuclei.
- Strangeness is longevity conferred on a transitional vortex of energy by a stable natural vortex at its core.
- Particles of matter and antimatter are subatomic vortices with equal mass but opposite direction of vortex motion (sign of electric charge).
- Gravity is a vortex interaction between matter and antimatter through the smallest realms of space.

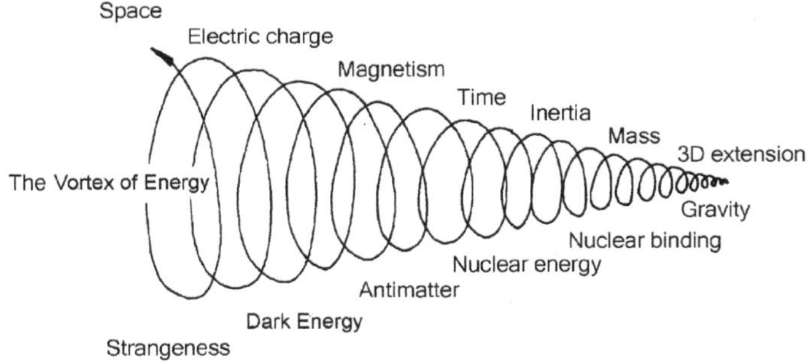

- Dark Energy is a vortex interaction between matter and antimatter through the largest realms of space.
- The Fourth Dimension of Space is the dimension of size of space.

The second requirement of a good scientific theory is to predict the outcome of future experiments, observations or discoveries. Prior to the 1990's it was assumed that the Universe was expanding at a uniform rate. However, in the 1990's Saul Perlmutter used redundant telescopes to look for supernova explosions in distant galaxies. In 1997 the results from observations of a couple of dozen supernovas were published which suggested the expansion of the Universe is speeding up.[4] Since then more supernova discoveries have supported this conclusion. Decades before Perlmutter's discovery it was a clear outcome of the cosmology of the vortex that the further galaxies are from us, the faster they accelerate away from us. The vortex cosmology was eventually published in 1995[5] two years before Perlmutter published his findings. By predicting the outcome of Perlmutter's observations before they were published, as well as satisfying Ockham's razor, the vortex theory satisfies the criteria of the scientific method for a sound scientific theory.

Science is not in the business of establishing truth. Science is in the business of establishing the believability of theories. If a theory satisfies the criteria of the scientific method it does not mean it is true but it can be incorporated in the body of science as being sound and believable.

No matter how successful a theory may be it will always be a theory; a story that we humans share to help us come to terms with the world in which we live. As Harari points out in his masterpiece, *Sapiens*[6] our success as a species is due, in large part, to the myths and stories we share. I believe this is especially true in the sphere of science.

Science is evidence based. But science is also based on a story; the story of materialism. Science can only improve on religion if there is a willingness amongst scientists to drop the philosophy of materialism in the face of incontrovertible evidence that it is unsound. Science is only credible if scientists treat all evidence with the same degree of impeccable impartiality. If evidence is

4 Perlmutter S. *Discovery of a Supernova Explosions.*Lawrence Berkeley National Laboratory, Dec 16, 1997

5 Ash D. *The New Science of the Spirit,* The College of Psychic Studies, 1995

6 Harari Y. *Sapiens: A Brief History of Mankind,* Vintage, 2014

rejected, because it doesn't fit the myth of materialism that most scientists happen to believe, that is prejudice and prejudice has no place in science. In science all systems of belief are hypotheses and theories which should be dropped immediately in the face of irrefutable contradictory evidence. If we lose sight of that fundamental rule in science we are liable to become lost and deluded. The time has come for this rule to be applied to the 19th Century theory of evolution.

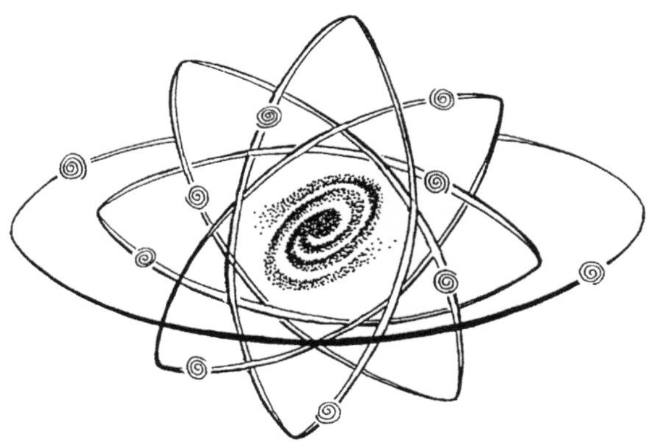

BOOK IV

THE ORIGIN OF LIFE

The old story has finally reached the end of its telling, and the space is clear for a new story to emerge. This cannot happen while the old story still carries hope.

Charles Eisenstein

CHAPTER 1
EVOLUTION IS NOT BY CHANCE

There is a problem with Darwin's theory of Evolution. The problem lies with the widely held belief that biological life, in particular the cell, appeared by pure blind chance from the random interactions of inorganic matter. That idea is not based on science, it is story that most scientists believe on the basis of their blind faith in materialism.

Charles Darwin

The chance hypothesis for the evolution of life on Earth caught hold in 1953 when Stanley Miller passed electrical discharges through a mixture of gases replicating the early atmosphere of the earth. Amino acids were discovered in the consequent mix which led the scientific community to conclude that Miller had discovered how life began. They seized on the idea he had proved that life emerged from lightening discharges which caused the formation of organic molecules from inorganic gases. In line with the predominant materialistic belief amongst scientists, the story that atoms jostled together to form amino acids, which then formed proteins morphed into 'scientific fact'. Over billions of years, scientists argued, these become organised, by incremental steps, into cells. This all happened by pure blind chance and the rest is 'evolutionary' history, so to speak.

The discovery of amino acids in meteorites supported this story. If life was seeded by the arrival of organic molecules from space, the possibility that life originated by chance was increased by the vastness of space, against the limits of earth.

However, the Miller-meteorite-mythology is weak. Science revealed another molecule, called RNA (ribose nucleic acid), as essential to the formation of proteins. RNA is made of small molecules called nucleotides, which are slightly more complex than amino acids.

In 2009 nucleotides were fabricated from electric discharges in a lab at Manchester University. However, this discovery did not explain how RNA first came to be formed. There is no evidence that RNA can form at random from nucleotides, just as there is no evidence that the appearance of amino acids, in Miller's experiment could lead to the happenchance appearance of a protein molecule.

Imagine we were wiped out in some terrible catastrophe and aliens landed on Earth outside a cottage in the Peak District. Unable to explain how it came to be they began to search for clues. One of them found stones on the ground identical to those in the cottage walls and suggested the stones must have arranged themselves into the cottage by chance.

The alien said that over an enormous period of time the random organization of stones could have happened incrementally, by the action of wind and wild animals. Another alien pointed out that the mortar, holding all the stone together in the cottage walls, was of the same substance as the stones. He wondered how it came to be set precisely between each stone and how they came to be perfectly positioned in the wall, with openings for windows and doors, just by chance. He concluded it was beyond the bounds of probability that both mortar and stones could have been arranged together, into the cottage walls, by luck alone.

The stones represent amino acids in this story and the mortar, the nucleotides. The random appearance of amino acids and nucleotides, through electrical discharges in gases, is the equivalent of the aliens finding limestone near the cottage. Just as the random appearance of limestone near the cottage does not explain the cottage, so the random formation of amino acid and nucleotide molecules do not explain how these came to be arranged in the protein and RNA molecules, observed in cell biology. The formation of organic molecules, in electric discharge experiments, does not support the belief that they self constructed by chance into complex molecules, in incremental lucky steps, over immense periods of time.

Protein molecules don't just self-construct. They consist of anywhere between fifty and several thousand amino acids organised in a very specific order, by a living cell, that enables them to perform specific tasks in the cell.

On average 200 types of amino acid are incorporated in a protein molecule. In every cell there are many thousands of different types of protein molecule. The chance of just one protein, consisting of 100 amino acids, randomly forming on earth, has been calculated at about one in a hundred thousand trillion.

The sequence of amino acids in a protein molecule is set by the RNA molecule. RNA is required to form protein but proteins are essential to the formation of RNA, so the proteins and RNA would have had to appear simultaneously perfectly formed for the first cell to come into being. The chance of that happening by pure blind chance is beyond the bounds of probability.

The alien, who came up with the chance hypothesis, went on to write a book called *The Blind Cottage Maker*. Just because the book attracted a large following amongst the people on his home planet it didn't necessarily prove it was true. It just proved that a lot of people don't think for themselves and can go along gullibly with what other people think rather than confront dilemmas in their own minds.

The chicken and egg dilemma of life is that new life always comes from existing life. Every cell originates from a cell that has existed before it. Cells divide to reproduce but the question is: How did the first cell come into existence?

All living cells fall into two basic types; a more primitive type called *prokaryotic* and a more advanced type called *eukaryotic*. Eukaryotic cells have a nucleus. Prokaryotic cells do not. Most bacteria are prokaryotic. Most of the cells in our bodies are eukaryotic.

Many biologists believe that the eukaryotic cell is formed of a number of different types of prokaryotic cells, which came together to form its organelles. That is a sound theory for the origin of the eukaryotic cell, which fits well with the theory of evolution. The question remains, however, where did the first prokaryotic cell come from?

The most primitive prokaryotic cell is an extraordinarily complex thing. Firstly it is surrounded with a permeable membrane made of protein molecules, which is many thousand times thinner than a sheet of paper but far more intricate. Think of the cell as a medieval walled city. Just as the walled city had guarded gates so the cell membrane also has gates formed of protein molecules that select what can enter and what can leave the cell.

Some protein gates are tubes that allow the passage of only very small molecules such as air and water. Others are open at one end and closed at the other, with the openings formed to fit a larger molecule of a specific shape. When a molecule with the right shape docks onto the opening, the gate closes around it to allow its passage across the membrane, while the other end of the gate opens to allow it to enter or exit the cell.

The inside of the primitive cell is full of watery liquid containing nutrients and waste, as well as useful end products of the chemistry going on in the cell. Just as the walled city is full of workshops, the prokaryotic cell is full of sites where very specific chemical reactions occur. Nothing is haphazard. Every reaction is ordered and scheduled according to the needs of even the most primitive cell.

Most of the production is concerned with the formation of proteins from twenty or so amino acids. These are used to build the structures of the cell, such as membranes, and the enzymes which facilitate the complex chemical reactions. The enzymes are arranged with proteins and RNA to form a ribosome. The ribosome is the chemical factory where the chemistry of the cell occurs. These are equivalent to the different workshops in the walled city, each designated to a different purpose i.e. the butcher, the baker or the candle maker.

Ribosomes make enzymes but they can only function with enzymes. How did the first ribosome form? The right amino acids required to form the right enzyme protein molecules must have come together, alongside an RNA molecule with the correct sequence of nucleotides. To imagine this occurred by pure blind chance is ludicrous because the math of probability does not allow for that degree of chance encounter. How could so many of the correct molecules come together and, entirely by luck, join in the required perfect, sequenced combination? It is a story that is hard to believe. We are left with the chicken and egg dilemma, which came first, the ribosome or the enzyme?

In both types of cell, instructions in messenger RNA are received from another complex molecule called DNA (desoxyribose nucleic acid), which occurs in a central nucleus in the eukaryotic cell. The instructions in the DNA molecule determine the sequence of nucleotides in the RNA and the sequence of amino acids in the proteins.

DNA is a double helix which coils to form the chromosomes in the nucleus of a cell. Even though prokaryotic cells don't have nuclei, they do have DNA double helix strands, delivering instructions from their binary code to RNA strands.

Another chicken and egg dilemma is raised by DNA. The formation

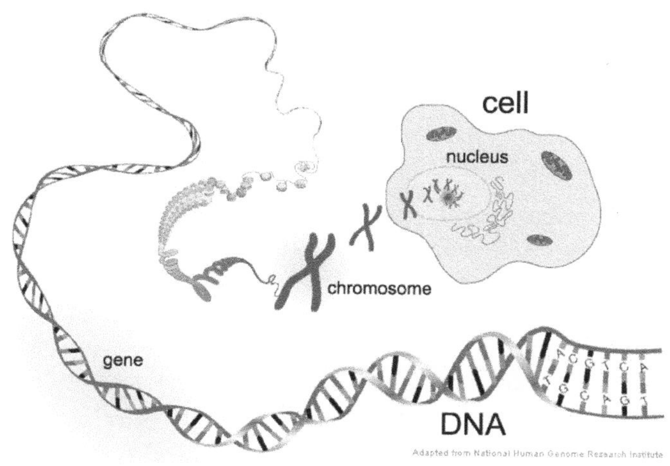

and replication of DNA requires RNA and protein enzymes but the formation of RNA and enzymes requires DNA. The first cell to form would require DNA with the right sequence of nucleic acids to instruct the RNA before the emergence of the protein enzymes. If this happened by chance there is still the problem of sequence to overcome. The enzymes, needed to form the DNA, require the DNA to instruct the sequence of amino acids required in their formation. Which came first, the DNA or the enzyme?

Finally there is the issue of energy. In the medieval walled city fire was essential for the baker to bake his bread, for the candle maker to melt his wax and for the housewives to cook the butcher's meat. But the fire had to be controlled in the fire place to give precisely the right amount of heat.

In the cell energy is required to enable all the many and complex biochemical reactions to occur. Energy in the cell is provided in precise amounts by ATP (*adenose triphosphate*) controlled within organelles called *mitochondria* which, in the cell, are the equivalent of the fireplaces in the medieval walled city.

Enzymes are needed to produce ATP but ATP is required to produce the enzymes. Which came first, the ATP or the enzymes?

Chicken and egg dilemmas abound in the cell; DNA cannot exist without enzymes but enzymes cannot exist without DNA. RNA needs proteins but proteins need RNA to form. Proteins are made in cells but cells are made of proteins. The simplest cell is made up of millions of proteins of thousands of different types. All these have to be working together, simultaneously, within a selectively permeable membrane for DNA, RNA and ATP to form and function.

Walled cities didn't suddenly appear on the landscape by chance. They were part of an evolutionary process involving intelligent people progressing from caves and mud huts. People with creative intelligence have been behind the evolution from mud huts to cities over the millennia and whereas many of the most important developments came about by chance, none of them occurred by chance alone. The existence of towns and cities is proof of the existence of intelligent people capable of designing and building them. In like manner biological cells are evidence of intelligence underlying the origin of life.

CHAPTER 2
LIFE SCREAMS INTELLIGENCE

In *The Intelligent Universe*[1] the cosmologist, Sir Fred Hoyle, reasoned that evolution is driven by some form of intelligence. Hoyle considered it a vast unlikelihood that life could have evolved from non-living matter without intelligence. In his words:

Sir Fred Hoyle

"...it is apparent that the origin of life is overwhelmingly a matter of arrangement, of ordering quite common atoms into very special structures and sequences. Whereas we learn in physics that non-living processes tend to destroy order intelligent control is particularly effective at producing order out of chaos. You might even say that intelligence shows itself most effectively in arranging things, exactly what the origin of life requires."

A process of evolution driven by intelligence is evident in our daily lives. We put out ideas and see if they work. If they don't we adapt them until they do, or drop them and try something different. Evolving theories are subjected to criticism so that only the best survive. But all the time we are in the process. The work would never happen without our conscious minds learning from the process of trial and error.

1 Hoyle F. *The Intelligent Universe*, Michael Joseph, 1983

Consciousness and intelligence are never excluded from the human creative process. Why would it be different for life in an intelligent Universe? As we take rejection in our stride and learn from our mistakes, we modify and make choices. There are lucky breaks but for the most part it is a learning process of step by meticulous step. In *The Blind Watchmaker*,[2] Richard Dawkins described the process of evolution, occurring as a series of incremental steps. That description doesn't preclude intelligence.

If, as Hoyle suggests, intelligence is a feature of the Universe, life could be the culmination of evolving intelligence on a universal scale. Maybe we are asking the wrong question. Debating whether or not life is a product of an independent intelligent being would be a pointless exercise if life were an expression of the ubiquitous intelligence of an intelligent Universe.

Life screams intelligence. Research is revealing the innate intelligence of plants and animals. Slime mould smart as a Japanese engineer and apes solving mathematical solution faster than humans suggests it might be a bit arrogant to exclude intelligence from evolution. There is no doubt Darwin's theory of evolution is brilliant and neo-Darwinian thinkers like Richard Dawkins have made contributions but perhaps it is time to allow for some form of intelligence in the theory of evolution.

Evolution makes sense as a means whereby Universal creative intelligence could have developed the multitude of forms that populate life. An intelligent Universe would never stop evolving, learning and growing, breaking things down and then building them up again better than before.

The extraordinary diversity and dazzling beauty we witness in life doesn't require an external creator that made everything perfect to begin with. The evolution of life on earth could be an expression of the intrinsic creative nature of the Universe. The Universe could be endlessly self creating through a process of intelligent evolution.

2 Dawkins R. *The Blind Watchmaker*, Norton & Co, 1986

Charles Darwin may have discovered a process by which Universal creative intelligence develops life. He didn't discover there is no Universal creative intelligence underlying life. One of Darwin's most brilliant insights was natural selection. Natural selection reveals mistakes. It is never easy for an author to spot errors. Authors rely on editors and critics to point out the flaws in their work. Life is harsher. Any organism that is not up to par ends up as lunch.

Chance is vital in evolution. Random events are essential to creative evolution because they enable freedom. Creativity is enhanced by freedom. I think Einstein was wrong when he said "God doesn't play dice." I believe it is through dice that the intelligent Universe plays. The Western mind fails to see the underlying order in apparent chaos and the perfect operation of chance. Not so the Chinese mind. In the words of Carl Jung, from a foreword to the *I Ching*:[3]

"The Chinese mind, as I see it at work in the I Ching, seems to be exclusively preoccupied with the chance aspect of events. What we call coincidence seems to be the chief concern of this peculiar mind and what we worship as causality passes almost unnoticed."

For intelligent systems random events and unforeseen problems create opportunities for creativity and originality. Evolution may be witness to Universal intelligence operating in biological systems, building things by trial and error, watching them perform then breaking them down in order to reconstruct them in a better way. This is the eternal cycle of creation, preservation and destruction – birth, life, death and then rebirth – depicted in the mysticism of India as the trinity of Creator, Preserver and Destroyer. People in the East do not preclude chance from their creation myths; they see chance as the footsteps of the divine.

In the chaos of random molecular interactions chance may have a vital part to play in the evolution of life but it is not necessarily the driver of evolution. Fervent belief in pure blind

3 Wilhelm R. *I Ching*, Routledge & Kegan Paul, 1951

chance, as the sole driver of evolution, is a lingering legacy of classical materialism in denial of the divine.

The 'blind chance' hypothesis, underlying evolutionary theory, originated in ancient Greek civilisation. It came from the atomic hypothesis of Democritus. Democritus taught that everything in the Universe comes from chance encounters of indestructible material atoms. The idea, that the Universe is essentially granular, came from his teacher Leucippus. It was based on the idea that everything is formed purely from the random interactions of bits of material substance. This is the underlying philosophy of materialism. It became the bedrock of science and, despite all the evidence to the contrary in quantum physics, it is still believed by many scientists today.

Many scientists turn their backs on scientific facts in order to harbour faith in materialism. In this regard scientists are no better than religionists in the adherence to blind faith. This is especially true in regard to the theory of evolution.

To date scientists, worldwide, have unearthed and catalogued some 200 million large fossils and billions of small fossils. Most researchers agree that this vast and detailed record shows that the major groups of all animals appeared suddenly and remained virtually unchanged. There is little evidence in the fossil record to support the theory of evolution. Even in *The Triumph of Evolution and the Failure of Creationism*[4] Niles Eldredge admitted that over time little or no evolutionary change in species appears to have occurred.

Though I am a fervent believer in evolution and natural selection, I have to admit the evidence for natural selection is very weak. In *Sudden Origins – Fossils, Genes and the Emergence of Species*[5] Jeffrey Schwartz wrote:

4 Eldredge N. *The Triumph of Evolution and the Failure of Creationism*, Freeman & Co., 2000

5 Schwartz J. *Sudden Origins – Fossils, Genes and the Emergence of Species*, John Wiley & Sons , 1999

"Natural selection may be helping species adapt to changing demands of existence, but it is not creating anything new."

The theory that mutations drive evolution has also failed dismally. In *Mutation, Breeding, Evolution and the Law of Recurrent Variation,*[6] Wolf-Ekkehard Lönnig, (of the Max Planck Institute for Plant Breeding Research) wrote:

"Mutations cannot transform an original species of plant or animal into an entirely new one. This conclusion agrees with all the experiences and results of mutation research of the 20th century taken together as well as with the laws of probability...properly defined species have real boundaries that cannot be abolished or transgressed by accidental mutations."

So why do scientists promote the theory of evolution in schools, colleges and universities as scientific truth when it lacks supporting scientific evidence. Influential evolutionary biologist, Richard Lowentin, said this is because, "...most scientists have a commitment to materialism and refuse to even consider the possibility of an intelligent designer."

The ongoing battle between evolutionists and creationists is a conflict of faiths. Neither camp is secure in experimental evidence. Both rely mainly on conjecture. We could walk away from this battle of beliefs and look at the scientific evidence in an unbiased way. The evidence of science reveals that everything is energy and energy is no-thing. Energy appears to be more akin to the nature of thought than material things. From this perspective, the Universe would seem more like a vast mental construct than material reality and so it would be ridiculous to exclude intelligence from evolution if nothing exists but thought.

6 Lönnig W. *Mutation, Breeding, Evolution and the Law of Recurrent Variation,* Recent Ressearch. Development. Genetic. Breeding, 2(2005)

CHAPTER 3
THE UNIVERSE AS A MIND

In *The Mysterious Universe,*[1] astro-physicist, Sir James Jeans wrote:

"Today there is a wide measure of agreement that the stream of knowledge is heading toward a non-mechanical reality; the universe begins to look more like a great thought than a great machine. Mind no longer appears as an accidental intruder into the realm of matter; we are beginning to suspect that we ought rather to hail it as the creator and governor of the realm of matter."

Many quantum physicists now see the Universe more as a mind than a material construct. If the Universe is a mind we would not need to imagine an intelligent designer to account for life. Mind and thought are synonymous with intelligence. If consciousness and intelligence are the very fabric of the Universe, there would be no need for a creator but it would also be illogical to exclude intelligence from the evolution of life. Both the creationists and evolutionists could be wrong.

Theologians imagined a God, existing in an eternity beyond the Universe, as the creator of the Universe and the life within it. However if the Universe is perceived as an intelligent, thinking, creative reality then there would be no need to imagine a supreme creator existing beyond it. The Universe could be self-creative and self-sustaining. This is a different story to the traditional religious one and it could satisfy the requirement, so many people have, for some sort of creative intelligence behind the origin of life.

A new story for creation began when Albert Einstein turned classical scientific materialism on its head with the equation

1 Jeans J. *The Mysterious Universe,* Cambridge University Press, 1930

$E=mc^2$. This equation shows the movement we call energy (E) at the speed of light (c) is the basis of mass (m). Einstein revealed atoms in matter are formed of particles of energy, not particles of material substance, and that was why he was never properly understood. In *Fields of Force*,[2] William Berkson said Einstein was difficult to understand:

"...not because of his ideas or the mathematics he employed, but because of his world view. Einstein denied the substantiality of matter and the field (light), whilst maintaining their reality."

Einstein discovered the non-material, non-substantial nature of the Universe. Energy exists but no-thing exists that is energy. Energy is no-thing. Energy is pure movement but nothing exists that moves.

Einstein realised that energy – in the form of particles of movement at the speed of light – underlies everything including mass, space and time. Energy is a measure of movement. Energy is not a thing that moves. Energy is the movement underlying all things and the Hiroshima bomb revealed that energy is not only the creator of things, it can also be the destroyer of things too.

How is it possible that particles of energy, which seem to be as ephemeral as thoughts, can form massive planets and blazing stars, solid mountains and extraordinarily complex living organisms like you and me? The answer to these questions can be found, not in the religious or materialistic philosophies of the West, but in the mystical insights of the East. The answer to the greatest enigma in modern physics, how energy can form and destroy matter, was discovered by Yogis in ancient times. The knowledge of the subatomic vortex of energy has been in the Yogic philosophy of India for millennia.

Long before Albert Einstein and the advent of nuclear physics, Yogis saw, quite literally, how energy forms mass. Millennia before the first atom bomb exploded over Hiroshima

2 Berkson W. *Fields of Force: The Development of a world view from Faraday to Einstein*, John Wiley & Sons, 1974

and the philosophy of materialism was blown apart, Yogis saw light spinning in the atom to form matter. They realised thousands of years ago that humanity has been deluded by spin to imagine that material things are real.

Only when we let go of materialism and our attachment to material things will we discover how we and the origin of life are inextricably linked. We are alive and conscious and the 'origin of life' may be the same creative consciousness as the consciousness that is in us, and that is also in everything and everyone else. If there is a God that is responsible for the origin of life that God could be the One Consciousness that underlies the Universe and all life within it, including you and me. In this view there would be no division between God and the Universe, no separation between God and humanity, and no dichotomy between creation and evolution. Yoga means *union*. Through yoga we can experience the origin of life directly and simply as the life force within us that sustains us all.

CHAPTER 4
CONSCIOUSNESS

You have just woken from sleep and look into the mirror. The first thought you have is, "Oh what a mess I am in this morning". The great debate is who is thinking that thought. Is it the body thinking about itself or is it an individual *psyche* in the body thinking about how the body vehicle looks.

A materialist would argue it is the body thinking about itself through atomic interactions in the brain. They would say that mind is just a manifestation of matter. It is the result of complex biochemical reactions in nerve cells.

A mystic might suggest the mind is thinking through the body about the body. They might say the mind exists separate from matter and incarnating in a body, mind expresses as thought in the realm of matter, through nerve cells, by means of complex biochemical reactions.

After breakfast you go out to the car. You look in the rear view mirror and think "Oh what a mess my car is in this morning." Would you say the thought about the car was coming from someone in the car thinking about the car, or would you say the thought was coming from the car itself?

We are so integrated with the body we identify with it as though we are it. But we may be a mind that moves in and out of the body like a driver moves in and out of a car.

When you are in the car you are an integral part of the car. It cannot drive without you. However, when you are in your car thinking, you do not imagine it is your car thinking. If you killed someone by dangerous driving the judge would not send your car to prison, no matter how vehemently you might argue it was your thoughtless car that was responsible.

Consciousness is the 'hard' question. Does consciousness originate in the brain or does the brain originate from consciousness existing as a Universal reality. Max Planck, the originator of quantum theory held the latter view. He said, "I regard consciousness as fundamental. I regard matter as a derivative of consciousness. We cannot get behind consciousness."

The debate about mind and body belongs in the realm of philosophy and metaphysics not science. It can be traced to the ancient Greek civilisation, to dialogues between philosophers like Democritus and Plato. Democritus, the father of materialism, argued that thoughts arise from atomic interactions going on in the body. Plato held the position that thoughts arise from a psyche that incarnates into the body before birth and vacates it at death.

Science does not take either side in this debate. If scientists and sceptics take the position of Democritus they do so not in the name of science but as individuals in the Western tradition preferring the philosophy of Democritus to that of Plato. Philosophy is not to be confused with science. The role of science is to provide evidence for people to examine so they can make a considered judgment of which philosophy, or story, they prefer to believe.

However, it is becoming increasingly difficult to keep the psychic out of science when, in the context of quantum physics, energy appears to be more in the nature of thought (Plato's psyche) than material substance (Democritus' atoms).

With energy appearing to consist of particles of movement, where nothing substantial exists to move, discussion on the nature of energy is impossible if psychic phenomena is ruled out of physics. The vortex theory crosses the traditional boundaries between physics and the non physical. As soon as the nature of energy comes into discussion mysticism becomes relevant because the Universe seems to be more akin to mystical than material reality.

In the wake of the quantum revolution science can no longer deal only with the material. If everything is energy and energy is non material, science can only survive by embracing the non material. Science will die unless the manifest non material nature of energy is accepted by scientists. Awaken to the non material reality of everything. Materialism is dead. It is materialism not God that is the delusion.

The vortex theory gives credence to a non-material view of reality. The vortex theory is upheld by the idea there are no-things as such. There is only the illusion of things set up by the spin in subatomic vortices of energy.

In the vortex theory energy is defined as movement where no-thing exists. Energy therefore seems more akin to mind than material substance. The understanding that a particle of energy is more like a thought than a thing presents a whole new view of the world. Where it was once considered unreasonable to contemplate spiritual or psychic phenomena in a scientific context, we now have a new scientific line of enquiry into this domain which does not attract scientific censure.

If particles of energy are more like thoughts than things then, as a body of energy, the Universe would be a mind. But thoughts depend on consciousness, so consciousness rather than material substance could be the bedrock of reality. In *The Self-Aware Universe,*[1] the Indian physicist Amit Goswami concluded that consciousness is the ground of all being. He said, "You can call it God if you want, but you don't have to. Quantum consciousness will do."

Goswami's view is shared by many quantum theorists but not the scientific establishment as a whole because mainstream science is still deeply embedded in materialism, based on the outmoded atomic hypothesis of Democritus.

Richard Feynman summarised the consensus view of the scientific community when he said:[2]

"If in some cataclysm, all of scientific knowledge were to be destroyed, and only one sentence passed on to the next generation of creatures, what statement would contain the most information in the fewest words? I believe it is the atomic

1 Goswami A. *The Self-Aware Universe: How Consciousness Creates the Material World*, Tarcher Pedigree, 1993

2 Calder Nigel, *Key to the Universe: A Report on the New Physics* (1977), BBC Publications.

hypothesis... that all things are made of atoms – little particles that move around in perpetual motion."

Quantum physics exposes the error in this materialistic world view. Particles of energy are particles of perpetual motion, rather than particles in perpetual motion. Even the greatest quantum physicists are challenged to understand the fundamental nature of energy being bits of movement rather than the movement of bits. Because of the misleading philosophy of materialism – that contends things rather than motion exist as prime reality – most scientists and philosophers are also challenged by the idea that particles of energy may be acts of consciousness rather than acts of material substance.

Appreciating subatomic particles as particles of vortex motion could explain how energy as pure, unsubstantial movement could give rise to particles of matter that exist as acts of consciousness. This may be the starting point of viewing consciousness as the universal creative principle.

In science and philosophy, consciousness is considered to be 'the hard problem'. Consciousness is a hard problem only because consciousness doesn't arise from anything; everything arises from consciousness.

Protons give us a clue to the nature of consciousness as a universal phenomenon. A proton is a particle of energy. Every proton that has ever been observed has precisely the same mass, the same quantum spin, the same electric charge and magnetism and gravity as every other proton. If protons are acts of consciousness these identical properties suggest they originate from the same consciousness. As protons constitute most of the mass of the Universe it would be reasonable to conclude that there is one source consciousness underlying everything.

We tend to think of ourselves as many conscious beings but we may be one conscious being in many bodies. If this is so, then if we hurt others we hurt ourselves and if we love others we love ourselves. From this perspective the principle of the brotherhood and sisterhood of humankind would be supported by an understanding of consciousness.

If a single consciousness underlies everything then this unity of being would also apply to plants and animals and even inanimate things like rocks, the soil and the sea. If this is so, one implication of understanding consciousness would be that if we are cruel to the planet and to plants and animals we are ultimatley being cruel to ourselves, but if we care for plants, animals and the planet we are actually caring for ourselves.

If particles of energy are more akin to thoughts than things, if they are more like acts of imagination than substantial material particles, then mythology could move to a whole new level of importance. It could be argued that every particle of energy, as an act of imagination, is closer to the nature of a myth than a material thing.

In science, energy is not properly understood. Physicists work with energy but they avoid the metaphysical question of *what is energy* by saying they don't know what it is. Before we can begin to comprehend quantum reality and Universal consciousness we need to understand energy.

CHAPTER 5
UNDERSTANDING ENERGY

A cosmic ray particle burst out of the sun with colossal kinetic energy. Screaming through space, close to the speed of light, it was only a matter of minutes before it entered the atmosphere of the Earth where a photographic plate – placed in a high altitude weather balloon by Professor Cecil Powell of Bristol University – was waiting for it. It could easily have passed right through the silver atom in the photographic emulsion, as it was mainly space, but with a bang it smashed into the nucleus of the atom. It hit a nuclear bull's eye.

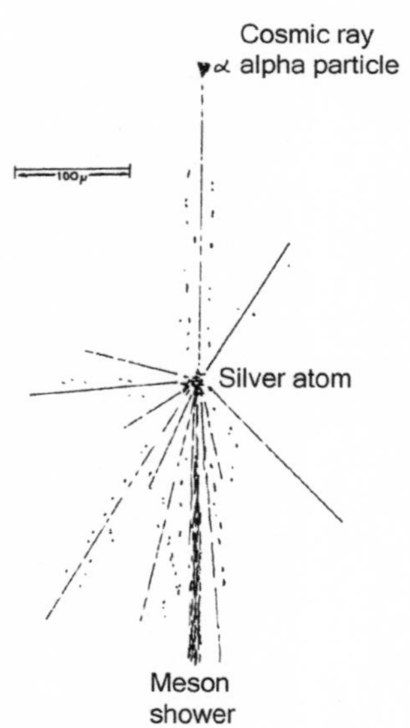

After the crash, the cosmic ray particle was no longer moving. Its progress was arrested by the nucleus of the silver atom as though it had hit a brick wall. But its kinetic energy went on. The *motion* of the cosmic ray particle left the pulverized particle behind and plunged on through the nucleus. Then the amazing happened. The *movement* of the cosmic ray particle exploded out of the nucleus, not as heat or light, but as a litter of new particles of matter. Energy as pure motion was transformed into mass.

The implications of this experiment are staggering. Plato taught that the human body is animated by a psyche; an energy that comes into the body at conception and continues on to a new life after the body dies. This core spiritual belief was rejected by Democritus, who taught there is no psyche or animating principle that survives death. He introduced the philosophy of materialism in which the body is considered to be animated by the activity of its atoms and molecules. According to materialism this animation cannot exist without atoms because animation is a property of the atoms. When the body dies the animation ceases. The atoms then disperse and may become incorporated into a new body. This materialistic belief is the creed of science.

Until the advent of nuclear physics, materialism ruled the day. Scientists believed that the movement or energy of a body could not exist apart from the body. Without the body the movement of the body ceased. It is obvious. It you are leaping around and then stop, the leaping stops too. The leaping doesn't go on leaping without you.

However, cosmic ray research and subsequent high energy physics has established that the leaping of high energy particles can survive head on collisions and the leaping can leap on, without the crashed particles, to form brand new particles of matter. The cosmic ray experiment established that not only can the 'animation' of a body exist without the body; it can go on to make a new body which is the core idea in reincarnation. Energy, traditionally defined as a property of material particles, turns out to be the essence of the particles themselves.

High energy physics does not appear to support the material hypothesis of Leucippus and Democritus. Albert Einstein established, in $E=mc^2$, that the energy that animates things also forms things but this is very hard for most people to understand.

How can the movement of a body exist without the body? How can it then go onto form new bodies? And yet it does. That is precisely what happened in the cosmic ray experiment, as witnessed by Professor Powell, and now repeated every day in experiments at CERN, Fermilab and other high energy laboratories around the world.

The same energy which moves everything about is ultimately what everything is made of. We in the West have been so indoctrinated by the material hypothesis that we find this idea very hard to grasp. That is why I keep repeating the mantra: *particles of energy are more like thoughts than things.*

Einstein was amongst the first to realise the error of materialism. But most physicists, embedded in materialism don't understand what energy is. As Richard Feynman said, "It is important to understand that in physics today we have no idea what energy is."[1]

Richard Feynman

People in physics are not prepared to say what energy is because energy is downright mystical. Energy has no mass or substance. It is no-thing. Energy is movement when there is nothing moving. Energy is activity when nothing is acting.

Fritjof Capra summed up the *energy enigma* in *The Tao of Physics*[2] and *The Turning Point*,[3] when he expressed the Indian mystic thought that anticipated quantum thinking:

1 Berkson W., *Fields of Force: The Development of a world view from Faraday to Einstein*, John Wiley & Sons, 1974

2 Capra F., *The Tao of Physics*, Wildwood House, 1975

3 Capra F., *The Turning Point*, Fontana, 1983

"There is motion but there are, ultimately, no moving objects; there is activity but there are no actors; there are no dancers, there is only the dance...The Vedic seers...saw the world in terms of flow and change, and thus gave the idea of a cosmic order an essentially dynamic connotation... Shiva, the cosmic dancer, is perhaps the most perfect personification of the dynamic Universe... The general picture emerging from Hinduism is one of an organic, growing and rhythmical moving cosmos; of a Universe in which everything is fluid and ever changing, all static forms being maya, that is, existing only as illusory concepts."

The vortex understanding of energy, derived from the ancient mystic tradition of Yoga, enables us to appreciate how energy that has no mass or material substance can, through spin, create a seemingly substantial and massive world. The subatomic vortex, seen by Yogis in ancient times, enables us to appreciate how energy – which is essentially dynamic – can spin to set up a seemingly static state. Vortex energy can explain the static inertia of mass. When we realise that there is no material substance underlying the things around us that appear to be solid and real; only then will we understand what energy is.

The vortex of energy enables us to appreciate the ultimate mystery of mind over matter. The vortex model shows us how energy can set up mass. Once we appreciate – as did the Yogis – that everything material is essentially mind and all manifest things are maya, an illusion of forms, then we can begin to understand the energetic nature of everything.

The Yogic approach to physics exposes materialism as a fable. The dynamic of intelligence, which we call energy, can be seen instead to evolve into everything, not by pre-design but by the innate tendency of quantum intelligence toward creative discovery, art and innovation. Into this field of creative imagination arises a possibility that traditional ideas, associating spirituality with the origin of life, might actually be pointing towards quantum intelligence and universal consciousness. Before considering these ideas the difference between spirituality and religion needs to be clarified.

CHAPTER 6
SPIRITUALITY IS NOT RELIGION

Spirituality is not religion. Religions follow scriptural teachings and often become cults by exalting their founder; sometimes to divine status. Religions are usually sectarian and adhere strictly to doctrines of belief. Religions are usually intolerant of dissidents as well as other religions. Religious are often involved with money and politics and the concentration of power.

In renaissance Europe, science arose at a time when those who dissented from religious dogma could be tortured and killed. Science came at a time when free thinkers were reacting against religions that had a long history of intolerance, bigotry, persecution and indoctrination. It was only natural that the free thinking pioneers of science, during the renaissance, were more inclined to secularism rather than religion – especially considering how the fanatically religious Roman Emperor Theodosius ordered, in 390 CE the destruction of all records of science and knowledge from antiquity that didn't accord with Catholic doctrine. This didn't happen in traditional Indian society where the truly spiritual people were tolerant and non-violent and approached their spiritual beliefs and practices with a degree of pragmatism.

Unfortunately today, in the minds of many people, spirituality and religion are lumped together. As religions lean more toward fundamentalism, secularists have become disparaging towards spirituality as well as religion. But fundamentalism is not peculiar to religion. It can happen in science too. Materialism, as well as monotheism, is subject to fundamentalism because both are erroneous streams of thought that engender fear.

There is a rising tide of fundamentalism amongst people who call themselves *sceptics*. In science and religion there is little difference in the intolerant attitudes of evolutionists and creationists. Fundamentalism is not restricted to religion or science. It is a fanatical human trait also found in politics. Fundamentalists are fanatics who believe that they are right and other people are wrong and that can be about anything.

The problem with the secular reaction to religion is it does not remedy religious intolerance but increases it. In her book *Islam*[1] Karen Armstrong wrote:

"Every single fundamentalist movement that I have studied is convinced that the secular establishment is determined to wipe religion out."

Fundamentalism in religion is usually based on a rigid interpretation of scripture along with a personality cult centered on the founder of the religion. In his book, *Sacred Scriptures*,[2] Timothy Freke wrote:

"Fundamentalism, usually fueled by scriptural writings has led to many of the horrendous religious conflicts that have bedeviled history. The mystical understanding, often drawing on the same scriptural inspiration, can help overcome these tragic divisions."

We should not allow fanaticism in religion to be a reason to demonise religion. After castigating religions on the hypocrisy of holy wars, the founder and executive director of *The Sceptics Society*, Michael Shermer, wrote in his book, *How We Believe*:[3]

"However, for every one of these grand tragedies there are ten thousand acts of personal kindness and social good that go unreported...Religion, like all social institutions of such historical depth and cultural impact, cannot be reduced to an unambiguous good or evil."

1 Armstrong K., *Islam: A Short History*, Phoenix, 2001
2 Freke T. *Illustrated Book of Sacred Scriptures*, Godsfield, 1997
3 Shermer M. *How We Believe: Science, Skepticism, and the Search for God*, Holt Paperbacks, 2003

Spirituality is different to religion. Mystics, or people who are truly spiritual, are not fanatical or attracted to money or politics; if they are they are not truly spiritual. Mystics live according to their truth whereas many religious people just talk about it. Spirituality and mysticism are personal journeys where individuals follow their inner guidance. Spiritual movements are not based on doctrine but personal experience and they are usually non-sectarian.

The Indian mystical and spiritual tradition of Yoga was never sectarian. The Yogic way of following a spiritual teacher, through the practice of yoga and meditation, was never based on indoctrination. Yoga has always been self realisation and union.

In the spiritual understanding of India, tolerance has always been of paramount importance. In India, scriptures are intended to guide people on a personal journey of discovery. Theology was less important than living with consideration for others (excepting the caste system) and all living things; especially cows. In Yoga the discovery of the nature of the world was encouraged as well as discovery of the self. Spirituality and mysticism rooted in Yoga are closer to science than religion because they are based on direct experience rather than adherence to belief or doctrine.

Through Yogic introspection Indians realised that there is a single, supreme, all powerful conscious intelligence that is in us all and in everything. They also came to appreciate that this universal wisdom manifests on Earth in various forms to elevate and teach humanity. Hindus respect all religious sects as being diverse paths back to the same source. In the words of Henry David Thoreau, "In the great teaching of the Vedas there is no touch of sectarianism."[4]

Religions usually followed in the wake of a spiritual teacher who was very often opposed to the religion of his day. Religions have led to empires and conflict, competition and war in total contradiction to their original ethos. India, despite the

4 BAPS *Shri Swaminarayan Mandir*, Neasden, London.

imposition of empires, has managed to maintain spirituality and true humanity to a greater extent than any other culture. Apollonius of Tyana, a contemporary of Christ, commented, "In India I found a race of mortals living upon the Earth but not adhering to it, inhabiting cities but not being fixed to them, possessing everything but possessed by nothing."[5]

What makes India so special is that spiritual exploration happened alongside the scientific. Yogic science went beyond spirituality and religion to discover, through introspection and personal development, that the Universal Consciousness in everyone also underlies everything. And the realisations of the Yogis helped in the understanding of quantum theory. As Werner Heisenberg said, "After conversations about Indian Philosophy some of the ideas of quantum physics that had seemed crazy suddenly made much more sense."

The scientific discoveries in the West may have taken our understanding of the Universe beyond that achieved by the people of ancient India but the underlying conclusions are the same. The Yogic perception, through *siddhi* powers, of the smallest particles of matter as vortices of energy, has enabled us to understand precisely how energy forms mass.

Thanks to the *Yogic vortex of prana* (energy) everyone can appreciate how energy can underlie everything. It is only one step on to appreciate that a single, unifying consciousness underlies energy. It is the coming together of ancient Yogic philosophy and $E=mc^2$, the single most important discovery in scientific history, that promises to unite us. It is the synthesis of science and spirituality that can cut through the many divergent beliefs that blight humanity.

In light of the insights passed down from India, the mother of science, there is reason to treat spirituality and mysticism with respect. Spirituality and mysticism are as old as humankind and we cannot dispense with them, in the name of science, and expect to survive. $E=mc^2$ is also the most terrifying scientific

5 BAPS *Shri Swaminarayan Mandir,* Neasden, London.

discovery in scientific history. Yoga can turn it into our salvation. Through Yoga we can appreciate energy as the innate universal spirit in everything.

As India can borrow from Western science and technology, so the West can borrow from the mysticism, philosophy, wisdom and tolerance of India. More than anything we need the Yogic understanding of the spiritual approach to humanity. As Arnold Toynbee said, "It is already becoming clear that a chapter that had a Western beginning will have to have an Indian ending if it is not to end in the self-destruction of the human race."[6]

6 BAPS *Shri Swaminarayan Mandir*, Neasden, London

CHAPTER 7
GOD AND ENERGY

In ancient India the Yogis recognised there was an ultimate creative principle underlying all phenomena. They called it *Brahma.* In science the ultimate creative principle underlying all phenomena is called Energy. In Christianity the ultimate creative principle underlying all phenomena is called *God.*

Brahma and God are attributed with consciousness and intelligence but the difference between Brahma and God is that in Christian theology God is placed outside of the Universe whereas in Yoga, which means *union,* there is no separation. Brahma is not consideredto be separate from the Universe.

There is a commonality between the definitions of energy in science and God in religion which suggests that God and energy may be one and the same.

- Energy is neither created nor destroyed
- God is neither created nor destroyed
- Energy is everywhere
- God is everywhere
- Energy is in everything
- God is in everything
- Energy is all might
- God is almighty
- Everything comes from energy
- Everything comes from God

In the Bible, the Gospel of John opens with the verse: "In the beginning was the Word and the Word was with God and the Word was God." Word is sound. Sound is vibration. Vibration is energy. So the first verse of John's Gospel could read: "In the beginning was the Energy and the Energy was with God and the

Energy was God." This translation of the first verse of the Gospel of John supports the idea of the oneness of God and Energy.

However, while God and Energy may be in union they may not be one and the same. This consideration comes from the idea that particles of Energy are more like thoughts than things. If the Universal Consciousness underlying everything is the thinker and particles of Energy are the thoughts then the thoughts would be in union with the thinker even though they would not be the thinker.

Thoughts are not things. Thoughts have no material substance. Thoughts are abstractions, they are acts of imagination. However, if no-thing exists the thinker would be as materially unsubstantial as the thought. The thinker would exist by virtue of thinking. It would not be a 'being' thinking, it would be a state of thinking; a capacity to think. If such a thinker ceased to think it would cease to be, just as the thought would cease to be. Thought may depend on a thinker for its existence but such a thinker would also depend on thought for its existence.

Thinker and thought would be in a state of total codependency. They would not be separate. They would be in yoga or union; two aspects of the same source of reality. The thinker would be the internalisation, and the thought would be the externalisation of reality, it would be the 'being' or state of 'expression' of the thinker.

If Universal Consciousness is the thinker, as well as thinking, it would have the capacity for conscious awareness. If Universal Consciousness is called 'God' then God would not be a being with a capacity to think and be aware, God would be the 'capacity' to think and be aware in all beings.

This is a very potent idea. We are conscious and we have the capacity to think and be aware. We are not God. We are mortal men and women but our conscious awareness and our capacity to think may be the principle we call God. We may be individuals with the capacities attributed to God. These capacities may go beyond consciousness and thinking to include intent, love, life, reason, creativity and many other attributes associated with the infinite mystery we call God.

With the abandonment of materialism, instead of imagining God as a being lording over us, we could imagine God as divine attributes associated with us and other sentient beings. We could think of God as what we are internally and the body as what we are externally. Maybe that is why the Bible tells us: *Be still and know that I AM God* (Psalm 46:10). Is God our I AM presence? Is God the essence of who we really are?

With the *Divine* capacities coming from within us, it would make sense to approach the Divine or God by going within. This is the way of the mystic. This is the way of Yoga. In the Gospel of St. Thomas, Jesus said, "If you don't go within, you go without."

As human beings we epitomise this relationship between God and energy. What is internal to us, our conscious awareness and our ability to think, create, love and reason, and our intent or free will, would be our inner Divine aspect. What is external to us, our physical bodies and genetic predispositions, our personalities etc., would be energy in the form of man (and woman) our outer mortal aspect. Together, what is external and what is internal makes us the sentient human beings that we are. This is why we are called *human beings.* 'Hu' is an ancient Egyptian word for God or the Divine. Mystics in ancient Egypt understood us to be Divine beings in mortal bodies.

In the Copenhagen interpretation for quantum reality it was asserted that quantum reality exists because it is observed, i.e. Energy exists to the extent it is observed by conscious awareness. This presumption is compatible with the idea of a creative, universal consciousness – call it God, Brahma or Quantum Cosnciousness – holding the Universe of energy in being through ongoing observation. The idea of God or Brahma as the Universal Observer fits with quantum thinking; and that Observer may be observing the physical Universe through us. It could be that God knows the world we live in through us.

All of this is philosophy. No-body knows for sure if God or Brahma exists. We only know the Universe exists. We do not know if these 'creators' pre-existed the Universe. However, we can consider the philosophical implications of the creator idea. Imagine a creator, in the eternal void before the Universe began thinking, "Hmm I think I will create a Universe!" There are philosophical problems with that idea.

Space and time belong to the Universe of Energy. Properties belonging to the Universe cannot be attributed to anything that may have existed before the Universe began. Voids and beginnings are attributes of space and time. They can only occur in a Universe that already exists. The proposition that they occurred before the Universe existed, before space and time began, is a logical absurdity. The idea of God or Brahma being an eternal state of Universal conscious awareness and intent, without beginning or end, fits much better with physics.

The idea that God existed before the Universe began doesn't make sense in the vortex theory. This is because time is considered to measure the relationships between the particles of energy that make up the Universe. By the same token the Universe could not have 'begun' some thirteen and a half billion years ago. If there was a big bang, it must have been preceded by a big crunch.

Think about it, the protons that make up the Universe are trillions of years older than the Universe itself. Protons have been around for at least three hundred billion, trillion, trillion, years; and maybe a lot longer. How can vortices of energy that 'make up' the Universe be trillions of years older than the Universe itself? There are serious metaphysical problems in regard to believing in a 'God' that preceded the Universe or believing the Universe began with a 'big bang'.

These problems do not exist if we are in one of many phases of Universal expansion. Philosophers in Yoga speak of the breath of Brahma, referring to the periodic expansion and contraction of the Universe. If the Universe goes through cycles of expansion and contraction then those who don't like the idea of the big bang could be satisfied and those who don't like the idea of God could just believe in an intelligent Universe which was, is, and forever will be, eternal and without end.

Yoga is union. Underlying Yoga is the union of the inner reality of consciousness and the outer reality of the observed Universe. In Yoga there is no separation between Energy and consciousness. As my daughter Jessica (a yoga teacher) said, "Everything goes back to Yoga."

If we tread the path of Yoga and mysticism rather than established religion and scientific materialism, we will come to know there is no division between God or Universal Consciousness and Energy. If there is common ground between God, Energy, Consciousness, the Copenhagen interpretation of quantum reality and the first verse of St. John's Gospel and if this commonality comes through Yoga then Yoga could turn out to be the loom that weaves the seemingly disparate threads of science and spirituality into a tapestry of knowledge for a new era when conflict is replaced with peace and goodwill between all mankind.

In *The Vortex Theory*[1], derived from Yogic philosophy electricity is treated as the equivalent of subatomic sound because the longitudinal vibrations of electricity are the same as the longitudinal vibrations of sound. If electricity is fundamental to life and if the silent sound of electricity were responsible for the Word spoken of in St. John's Gospel, then creative expression of thought and the transmission of information throughout the Universe as electric impulses of life would be another way of translating those most profound words in the Bible "... as the electricity was with God and electricity was God."

If there is a connection between God and electricity, and if as many people believe God is the origin of life, it may be worth considering the connection between electricity and the origin of life.

1 Ash D, *The Vortex Theory*, Kima Global Publishers 2015

CHAPTER 8
PLASMA AND LIFE

Physics allows for the possibility of other forms of matter apart from atoms. It is this possibility that opens a window of opportunity to expand our understanding of life and its origins.

Ninety nine point nine percent (99.9%) of the Universe consists of non-atomic matter. In non-atomic matter vortex particles are not organised into atoms. Non-atomic matter is sometimes called *plasma*.

Only 0.1% of the mass of the Universe exists as atoms of matter. These atoms are not billiard balls of material substance, as envisaged in classical physics. They are systems of subatomic particles which, according to the direct perception of Yogis, are vortices of energy.

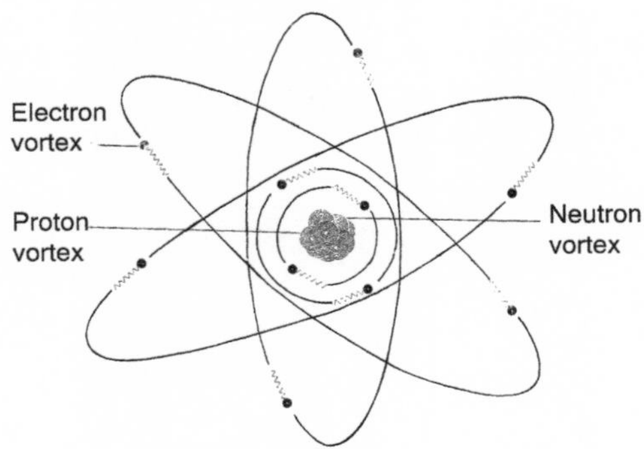

Most people are familiar with the picture of the atom with electrons orbiting a single, central nucleus. Applying the Yogic perception, the atom would be electron vortices of energy orbiting proton and neutron vortices in a central nucleus. The 'dynamic' vortex energy, extending beyond our perception, would be responsible for the force of electric charge acting at a distance from the matter we perceive.

In the atom, opposite charged particles are paired off so that the atom is electrically neutral. If life depends on electricity, atomic matter would be the worst stuff in the Universe for supporting life. If electrons are added to an atom or taken away from an atom, then the atom becomes an electrically active *ion*. In atomic matter ions can support life because they can conduct electricity.

Plasma is formed of ions therefore plasma would be good for generating the electric conditions favourable for life. This is because plasma is electrically active. If most of the Universe is plasma and ideal for the electric conditions of life, the Universe could be brimming with life, in fact a plasmic Universe could be alive; it could be the equivalent of a living organism. Out in space there could be forms of life, sentient beings even, which like us could embody attributes of Universal Consciousness such as intent and the ability to think, love and reason. But they would be invisible to us because cold plasma in space is invisible. This is because cold plasma does not necessarily emit or reflect light.

Invisible plasmic life-forms in space, if they exist, could be called spirits. This is because plasma in space is very low density and the traditional folk term to describe things that are low density and invisible is *spirit.*

If the nucleus of an atom were the size of a golf ball, the nearest orbiting electron would be about two miles away. In plasma, the electrons and atomic nuclei would be much further apart. In the sun, temperatures are too high for atoms to form. Most electrons are moving way too fast to stick around a single nucleus therefore the distance between them would be far greater than in an atom. 98% of the mass of the solar system is contained in the sun. This mass is plasma. The sun is formed of hot plasma that emits light. The planets orbiting the sun are made of atomic matter. Combined, the planets add up to a measly 2% of the mass of the entire solar system.

We see the moon and planets because the atoms in them reflect the light of the sun. We see things on earth due to light bouncing off atoms and ricocheting into our eyeballs. This is mostly reflected sunlight. Light doesn't reflect off plasma but we see the sun because its hot plasma emits light. If plasma isn't emitting light we don't see it. Cold plasma in space doesn't necessarily emit or reflect light. That is why it is invisible.

Space is full of cold plasma because stars release plasma into space as solar wind, cosmic radiation and solar flares. These charged particles in space could form spiritual bodies. Plasmic bodies could be the basis of non-biological, non-atomic life forms. These might eat plasmic matter to feed their bodies just as we eat atomic matter to feed ours.[1]

These plasmic or 'spiritual' beings, if they exist, could be aware of each other and their spatial environment by feeling the forces of electric charge – extending vortex energy – that lie between them. In spiritual *plasmic* bodies the electric 'vortex energy' fields would be more fluid than in the physical *atomic* bodies we inhabit. In atomic bodies electric fields act as local locking devices between atoms. This is very different to the more fluid like electric fields in non-atomic plasmic bodies. The difference between 'physical' and 'spiritual' life could be the difference between atomic matter and plasmic matter.

Once we start thinking of the spiritual in terms of plasmic matter we can embrace the spiritual in physics and include spiritual matters in science. Once we address life in terms of electrical activity and electric fields then we can begin to study the links between physical and spiritual life. A line of enquiry could be a programme of research into the possibility that there may be life in space, based on cold plasma.

Despite the logic that there might be extraterrestrial life in space, formed of plasma that the Universe is mostly made of, it is unlikely that such a line of enquiry would make much headway in the current university establishment. The cult of materialism is too entrenched in secular society. It would be like trying to get the Catholic Church in medieval Europe to consider Hinduism as a religious option. Taking into account the entrenchment of materialism in scientific orthodoxy it may be best to treat these non-material unorthodoxies as mythologies or stories until the

1 Heley Mark, *The Everything Guide,* Adams Media. 2009

climate of scientific opinion is more open to the possibility that spirit might be plasma.

The great value of Yuval Noah Harari's masterpiece *Sapiens: A Brief History of Mankind*[2] is the importance he places on the development of myths in the historical success of humanity. In his own words:

"Any large scale human cooperation – whether a modern state, a medieval church, an ancient city or an archaic tribe – is rooted in common myths that exist only in people's collective imaginations…Yet none of these things exists outside the stories that people invent and tell one another."

In this context we can develop a new mythology or story based on the idea of spiritual 'space beings' being non-atomic plasmic forms of life. Russia is well in advance of the West in the scientific investigation of links between plasma and life in outer space so this is not an unreasonable proposition.[3] If plasma research were taken as seriously in the West, as it is in Russia, then this might be considered forward thinking and might be the basis of a new scientific story.

Scientific stories are called world views or *paradigms*. World views help us to invent hypotheses that can be developed into a scientific theory. Rather than 'believe' in them we could consider with impartiality the possibilities they represent.

The key to understanding how plasmic beings could be involved in the origin and evolution of life on Earth lies in water. If electricity is vital to plasmic life then a medium in atomic matter, as near as possible an equivalent to plasma, would be necessary to support the conductivity of electricity essential to life. Salty water conducts electricity. Salt water on Earth is a fluid medium capable of conducting electricity. Maybe salt water is acting in every living cell, as the equivalent of plasma in space.

2 Harari Y. *Sapiens: A Brief History of Mankind*, Vintage, 2014

3 Heley Mark, *The Everything Guide* Adams Media 2009

CHAPTER 9
WATER AND LIFE

In biology we learn about the properties of living organisms like respiration, locomotion and reproduction etc. but the most basic property of living orgamisms, what makes them 'alive', is electrical activity. Electricity is the key to understanding life and the origin of life. This can help us appreciate why life on Earth originated in water and why water, especially salt water, is essential to life.

What is it about water that enables it to support life? The answer may lie in its ability to conduct electricity.

Electrical conductivity originates from charged particles. Most atoms are electrically neutral because they contain an equal number of oppositely charged particles. But they can become electrically active by gaining or losing electrons. When that occurs they become ions.

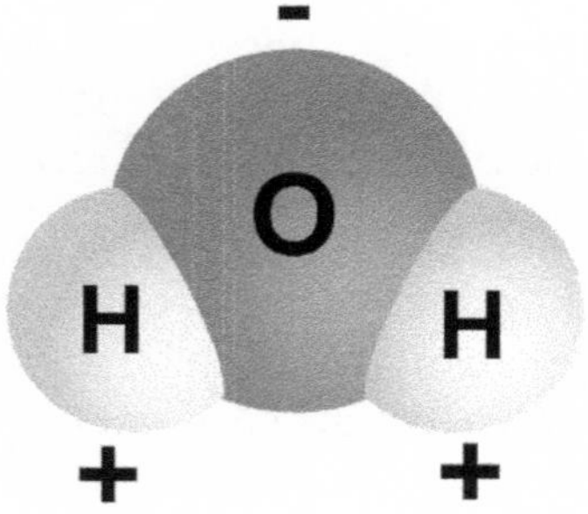

Water is formed when an oxygen atom is combined with two hydrogen atoms. In this combination the hydrogen electrons are predominantly in the orbit of the oxygen because oxygen atoms are electron hungry. The result is electrons are more likely to be found on the 'oxygen' side of the water molecule rather than where the hydrogen atoms are. The two hydrogen atoms then act as 'horns' of positive charge while the oxygen acts as a 'belly' of negative charge.

This *electric dipole* of water sets up a small electrostatic force of attraction between one water molecule and another, 'belly to horn' which enables a weak bond to occur between them. This is called *hydrogen bonding*.

The water molecule is not an ion but its electrostatic properties enable it to dissociate salts into ions. This occurs when salt dissolves in water. Salts, when they dissolve, release charged ions into the water. These enable water to conduct electricity. This could be why salt water is the basis of life on earth.

The biological cell is essentially a drop of salty water suspended in a jelly of protein. The salt in the sea is predominantly sodium chloride. But sea salt also contains magnesium ions and scores of other ionic minerals which are vital for the healthy functioning of cells.

Life depends on the electrical conductivity of salty water. Salty water, full of *electrolytes,* is called *plasma*. The fluidity and electrical activity of watery plasma in biological cells could be on Earth the nearest thing to life plasma in space.

If there is plasmic life in space it would make sense that a planet containing an abundance of watery plasma would be ideal for establishing life in atomic matter. This may be why biological life, based on atoms and molecules, first emerged in the salty sea.

The idea that the Earth has been colonized by biological life originating in space is not new. What is new is the idea that extraterrestrial forms of life may be unlike anything imagined in science or Sci-Fi. The forms of life responsible for the origin of life on earth may be invisible, low density, electric field generating space plasma, rather than little green men.

An association between electric fields and life was discovered by Harold Saxton Burr.

Harold S. Burr

In the 1930's, while he was professor of medicine at Yale University, Burr used a high impedance volt meter to measure electric fields in and around living organisms. He named these, Electrodynamic Life Fields or *L-fields*. He published around thirty papers on his extensive research in which he determined that the life field has a major impact on the process of differentiation of multicellular organisms.

In *Blueprint for Immortality*[1] Burr concluded:

"When a cook looks at a jelly mould she knows the shape of the jelly she will turn out of it. In much the same way, inspection with instruments of an L-field in its initial stage can reveal the future 'shape' or arrangement of the materials it will mould. When the L-field in a frog's egg, for instance, is examined electrically it is possible to show the future location of the frog's nervous system because the frog's L-field is a matrix which will determine the form which will develop from the egg."

The electrodynamic L-field, also called the *biofield,* has holographic properties in that any portion of the Life-field contains the blueprint of the organism. In the early stages of its development, if the cells of an embryo are divided in half, each half will develop into a completely formed organism. This occurs with identical twins. Just as the division of a hologram results in two complete images, the division of the embryo results in two identical twins. Like the hologram it seems every part of the L-field is a blueprint for the whole.

In *The Field,*[2] Lynne McTaggart cited a number of scientists, associated with universities in the USA and France, Germany and Russia, who discovered a close association of electric and magnetic fields with life. The question that goes with these discoveries is how a low density electric field could overlay and inter-penetrate a body of atomic matter. Very strong electric fields occur in atoms, between electrons orbiting protons in the

1 Burr H.S. *Blueprint for Immortality,* Neville Spearman, 1972

2 McTaggart L. *The Field,* Harper & Collins, 2001

nucleus. How can the electric fields of life penetrate the powerful electric fields of the atom?

Space plasma might influence atomic matter if it were operating in a way than did not involve a direct interaction with the electric fields in atoms and molecules. Developed from the vortex theory, a new paradigm or *frame of understanding* could render interactions between the electric fields of life in space plasma and molecules in biological cells plausible. This new frame of understanding rests on the axiom that there are levels of energy in the Universe based on speeds faster than the speed of light.

CHAPTER 10
SUPER-ENERGY

In his theories of relativity Albert Einstein established the importance of the speed of light in our world. Einstein said that mass, space and time are relative to the speed of light, which is measured at 299,792,458 metres per second.

Einstein declared the speed of light as the sole universal constant. He may have been wrong in that assumption. It may be that the speed of light, measured by scientists, happens to be the intrinsic speed of energy that makes up our world. It could be that in the Universe there are other worlds made up of energy based on different speeds of movement. This postulate is the basis of the new *paradigm;* a new *frame of understanding* for multidimensional life.

In the vortex theory it is contended that the substance of everything is movement rather than material substance. High energy physics reveals there is no-thing existing but particles of movement. In the Einstein world view there is no-thing that moves.[1] Einstein determined that only the movement we call energy exists. Energy is the substance of everything and everything is relative to its speed.

If our world of energy is made up of particles of movement at the speed of light there could be other worlds in the Universe made up of particles of movement defined by speeds faster or slower than the speed of light. If the movement underpinning our world is spinning in vortices and undulating in waves at the speed of light, then there could be other worlds, or levels in our world, in which the movement spinning in vortices or undulating in waves is faster than the speed of light.

1 Berkson W. *Fields of Force: The Development of a world view from Faraday to Einstein,* John Wiley & Sons, 1974

The movement that forms our world is called energy. The motion forming worlds beyond the speed of light could be called *super-energy*. The new paradigm, or frame of understanding, is based on this premise; *super-energy forms worlds beyond the speed of light and these worlds could be just like our own.*

We may need to revise our use of language. Nouns describe things and verbs define actions but in the new quantum physics, based on the vortex, there are no nouns, only verbs and the verbs have become nouns by spinning. There are no-things, as such, because energy, which is the basis of everything, is not a thing; it is pure movement. Things only appear to be real and solid because of spin.

The idea is that this principle can apply to energy based on speeds faster than the speed of light. Super-energy, faster than light, could spin to form subatomic particles of matter and undulate in waves to form photons of light. These would then form worlds, just like our own. This is the basis of the new *super-energy paradigm*, which enabled me to incorporate the *supernatural* in my new scientific frame of understanding.

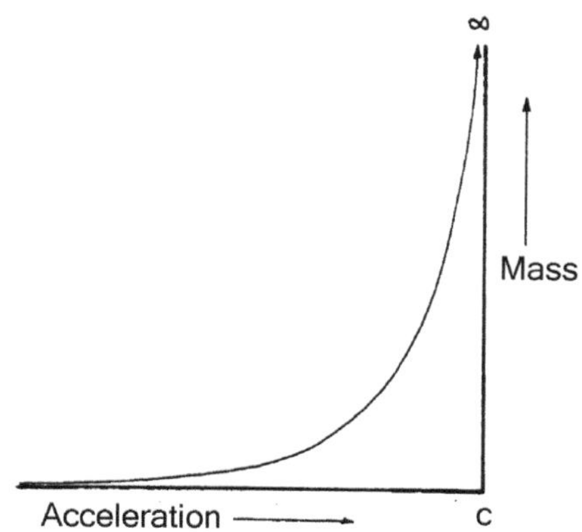

The idea that energy exists with speeds faster than that of common light does not contradict Einstein's theory of relativity. Relativity theory makes it clear that as particles accelerate they don't exceed the speed of light – represented by the symbol c – they just become ever more massive.

In the vortex theory it is obvious that a particle of matter can never move faster than the speed of the energy propelling it. In the vortex model the wave-trains of energy driving into the vortex particle that cause it to accelerate also increase its mass. This happens because energy can be captured by the vortex. The ever increasing mass of an accelerating vortex then requires even more energy to accelerate it to higher velocities. This exponential increase in the mass of particles of matter, with acceleration, was predicted by Albert Einstein and has been confirmed in particle accelerators. As explained in the vortex theory, the ability of subatomic vortices to capture energy accounts for why accelerating particles of matter, like electrons, are more massive than particles at rest.

However, high energy physics does *not prove* that the energy within the vortices and the waves propelling them, cannot exist somewhere in the Universe with an intrinsic speed faster than the speed of light, i.e. super energy. There could be worlds made up of vortices and waves of super-energy governed by the same laws of physics that underpin our world. In the new paradigm, the premise is that the only difference between our world and the worlds based on super-energy would be the *relativistic speeds* of movement underlying them. These 'speed of light' relativity constants could be called *The Einstein Constants* after Albert Einstein. This is in view of the fact he determined that the laws of physics in our world are relative to the speed of light (the speed of the energy that makes up our world).

The physical universe might be only a small part of the whole universe. The universe may not be restricted to the world we see with our senses – and observe scientifically through telescopes and microscopes. The world from minute atom to mighty galaxy might be merely a part of a much greater universe of energy. The universe might be vaster by far than science has previously allowed. The 'heavenly realms' may be real, a parallel reality created out of super-energy. Just as our world and the thiings within it are forms of energy , so this other reality could consist of forms of super-energy. We could not travel to the realms of super-energy through space and time because the fabric of space and time is set up by the vortex. Heaven would not be 'up there somewhere', heaven may be beyond the speed of light.

In a new paradigm, based on the idea of super-energy, it is postulated that there are worlds beyond our own and the next world is governed by an Einstein constant equal to about twice the speed of light. This world is called *the hyperphysical plane of reality* and would correspond to the fourth dimension in metaphysical parlance.

In this new paradigm the hyperphysical level of super-energy is used to account for the *soul.* The hyperphysical plane of reality is considered to be the world where souls reside. Souls, which are essentially individualisations of mind, could have *hyperphysical bodies* to live in.

In my scientific story a soul incarnate in its hyperphysical body can overlay and interpenetrate a physical body because of a *principle of simultaneous existence.* Because the vortex sets up space and time on each *plane* of reality there would be no space-time separation between the planes. The only separation between the physical, hyperphysical and super-physical levels in the Universe would be their Einstein constants; their *relativistic* intrinsic speeds of energy. Therefore they would all occur in simultaneous existence. The best way to imagine this is to liken each plane of reality to a sheet of paper on an office desk. Imagine the layered sheets *coinciding in the same here and now* by being impaled on a single spike.

In the new paradigm of super-energy, our soul bodies are made up of atoms, molecules and cells just like our bodies of physical matter. Soul bodies are made up of *hyperphysical matter.* The only difference between our physical bodies and our soul bodies would be the speed of the energy spinning in the vortices and undulating in the waves that make up these bodies. Souls could be living in hyperphysical bodies on a *hyperphysical Earth* just as we are living in physical bodies on a physical Earth. The hyperphysical Earth could be a level of super-energy, associated with the Earth. It could be the *Earth Soul.*

In the super-energy frame of understanding, there are worlds beyond the hyperphysical level which are thought to be governed by Einstein constants higher than twice the speed of light. These worlds are called the *superphysical planes of reality* and would correspond to the fifth dimension and beyond in metaphysical parlance. These super-physical levels of super-energy would be associated with deep space and correspond to *higher dimensions of reality* and also to the *Divine.*

I use a concentric sphere model to depict the physical, hyperphysical and superphysical levels of energy. This corresponds to the *Pythagorean Harmony of Spheres* that was used by philosophers in the classical period to describe the Earth and the heavens as concentric planes of reality. The fundamental premise of my new super-energy paradigm is that on all the planes there would be light and matter – both atomic and plasmic – and the laws of physics would be the same; only the underlying speeds of energy, the Einstein constants of relativity, would be different.

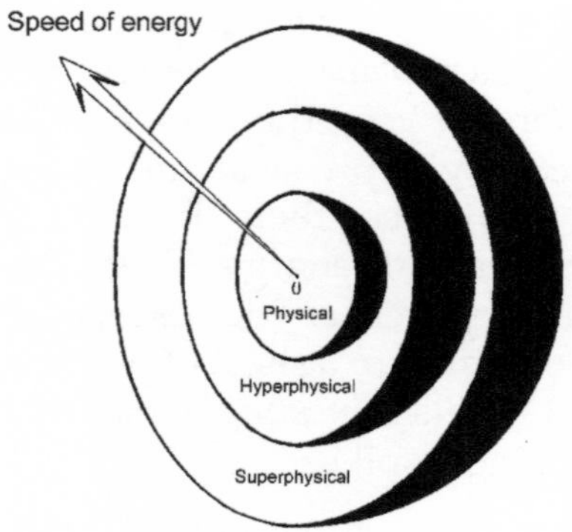

The physical world, based on the speed of light, occupies the central sphere. The hyperphysical world is represented by the second sphere and the super-physical is represented by the third sphere out. This arrangement illustrates the fact that *slow speed values are a sub-set of fast speeds,* i.e. the maximum speed of a bicycle is a sub-set of the speed of a car which in turn is a sub-set of the speed of a jet.

No matter how hard you pedal on your bike you won't catch an accelerating car. In the same way, a jet can move at the speed of a car but a car can't keep up with a jet. This is the basis of a principle which I call the *principle of energy speed subsets.* Slow speeds are a subset or a part of faster speeds and they cannot keep up with the faster speeds. This principle explains the relationship between the physical level, the hyperphysical level and the superphysical levels of energy in the Universe, as depicted by the model of concentric spheres; the *Harmony of Spheres.*

In *The Vortex: Key to Future Science,*[2] my co-author, Peter Hewitt, and I used a 'matchbox analogy', to illustrate the relationship between the planes of super-energy and the plane of energy that we live on. We depicted the physical world as a matchbox. We then imagined the matches in the match box as people that were only aware of the world inside the match box. They were not aware of the room outside of the matchbox or the folk in the room who are aware of them. In our analogy the room represented the hyperphysical level of reality. The room, in turn, was part of a house, which represented the superphysical level of reality.

You cannot fit a room into a matchbox and you cannot fit a house into a room. So it is that movements with speeds beyond the speed of light could not be contained or constrained in the physical space and time relative to the speed of light. This is an important implication of the principle of energy speed subsets. Seeing-is-believing does not apply when it comes to perceiving supernatural phenomena because of this all important consequence of the *principle of energy speed subsets.*

A super-energy being could be in your presence because your space would be part of its space. It could see you but you would not see it. This is because physical light would pass right through it. Physical energy does not interact directly with super-energy, so photons of physical light would not reflect off the super-energy of the supernatural being. That is why it would be invisible to you. It could see you because you and everything in your environment would be a part of its space-time continuum. This is why *seeing-is-believing* limits us. Things could be going on all around us which we cannot see with our physical eyes.

While there would be no direct interaction between bodies of super-energy and physical energy there is a possibility that an indirect action could occur between the planes of energy and super-energy through *resonance.*

2 Ash D. & Hewitt P. *The Vortex: Key to Future Science,* Gateway Books, 1990

CHAPTER 11
DNA RESONANCE

The work of science is to help humans understand the Universe and explain the mystery of their existence. The new paradigm of super-energy, based on the premise that there are worlds of energy existing beyond the physical speed of light, could help to shed light on the origin of life.

It's a credit to the human race that there are scientists like the renowned Cambridge biologist Rupert Sheldrake[1] who have the courage to attempt to explain phenomena beyond the physical in a scientific context. Appreciating there may be fields of energy beyond the speed of light interacting with our biological world could help in understanding some of his ideas, as he is responsible for introducing the idea of *resonance* into biological systems.

Rupert Sheldrake coined the term *morphic resonance* to describe the ability of living systems to inherit a collective memory from all previous forms of their kind. The word 'morphic' is derived from the Greek word *morphe* meaning 'form'. Reading Sheldrake's proposal for the possibility of resonance between 'like lifeforms' suggested to me and my coauthor, the late Peter Hewitt, that there may be a link between morphic resonance in biological systems and DNA.

Peter and I proposed a form of resonance involving DNA, which we called *DNA resonance*. In *The Vortex: Key to Future Science,*[2] we used our idea of DNA resonance to explain how the Life Field, proposed by Professor Harold Saxton Burr, could underlie the processes of differentiation and evolution.

1 Sheldrake R. *A New Science of Life,* Paladin Books, 1987

2 Ash D. & Hewitt P. *The Vortex: Key to Future Science,* Gateway Books, 1990

According to orthodox biology, differentiation is controlled in chromosomes of DNA by genes switching on or off. This is a process which involves complex chemical reactions but according to the research of Professor Burr, the underlying direction of differentiation appears also to be coordinated by an electric-field matrix overlying the cell, most especially the DNA molecule.

The proposal in the new paradigm of super-energy is that there are living, sentient and intelligent *superphysical plasmic fields* interacting through resonance with the coincident DNA molecules in hyperphysical and physical matter. This could operate in a similar way to how broadcast electromagnetic fields interact with radio and television sets by resonance.

The postulate, in the new paradigm of super-energy, is that plasmic fields in the superphysical levels of reality resonate first with hyperphysical DNA then the hyperphysical DNA resonates with physical DNA in biological cells.

DNA resonance through DNA molecules is considered to be a process of information transmission between the planes of super-energy and energy. This resonance would occur between the super-physical, the hyperphysical and the physical levels of reality. In the new paradigm, *DNA Resonance* is considered to be the key to the origin and evolution of life on Earth.

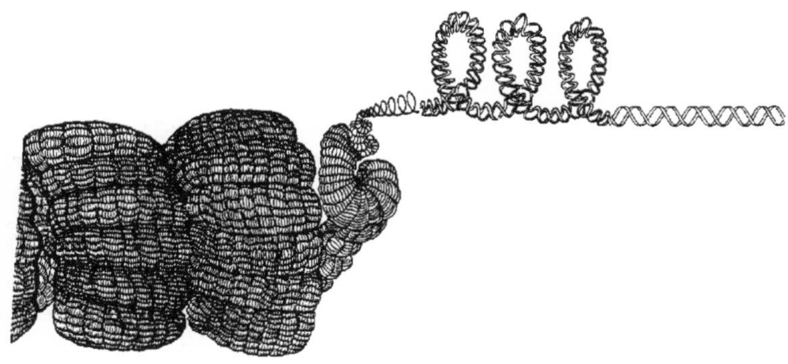

The DNA molecule is a double helix coiled upon itself several times. This structure is reminiscent of the coil in a radio or television set that enables it, through resonance, to receive a broadcast program of information. The coil structure of DNA may allow the life-fields, also known as *biofields*, to resonate with the DNA.

Biofields, or life-fields, have been recorded by a number of scientists, as well as Burr. These include a British doctor Walter J. Kilner, who invented a series of goggles and filters through which anyone can see biofield auras in detail and Harry Oldfield, who invented *Polycontrast Interference Photography* (PIP) to register patterns of light imperceptible to human eye that radiate from biological organisms. Biofields, as we perceive them, could be the result of a resonance between physical electric fields and the 'living electric fields of superphysical plasma'.

According to the new paradigm, DNA could be acting, through its structure, as a receiver of information, broadcast from higher levels of reality in the Universe. Super-energy, as a carrier of information, could be mediating through the planes by layered DNA resonance between the superphysical and the hyperphysical and then between the hyperphysical and the physical. Thus, via DNA resonance living, superphysical plasmic fields could be involved in the origin and evolution of life on earth.

A future line of enquiry in science could be to investigate the DNA molecule for signs of resonance to see if it responds to electric fields in line with the extraordinary research of people like Burr. The research could include repeating Burr's experiments by exposing cells to electric fields to see what impact they have on the process of differentiation.[3]

These questions have the potential to lead us to new vistas of scientific enquiry. Could a program of information coded with a recipe for terrestrial life stem from superphysical levels of reality? Could such a program of information be carried as field frequencies in a living plasmic electric field? Could this shed some light on the traditional idea of 'spirit' acting on the physical domain?

Maybe a resolution to the ongoing conflict between creationists and evolutionists could be found in the process of

3 Burr H.S. *Blueprint for Immortality*, Neville Spearman, 1972

DNA resonance. Maybe elements of both spiritual creation and terrestrial evolution belong in an account for the origin of life.

Codified information from the superphysical, passed into the hyperphysical DNA, could be transferred to a physical DNA molecule counterpart through resonance based on 'identicality of form'. It could be the matching forms of the hyperphysical DNA overlaying physical DNA that would allow for resonance to occur between them.

If a tuning fork tuned to 'A' is sounded, the 'A' string on a guitar will vibrate but the other strings will not. Only the string tuned to the frequency of the fork will respond or resonate to the tuning fork. This is because frequency is a measure of the shape or form of a vibration. When the guitar string is tuned to the tuning fork, its vibrations have the same matching *frequency form or shape.*

A key to the working of this model is the proposal that because there is no space-time separation between the three planes of reality, *the principle of simultaneous existence* could apply. The electric plasma field of the superphysical could overlay and coincide with the hyperphysical DNA, which could then overlie the physical DNA; in much the same way Russian dolls sit within each other.

The *principle of energy subsets* then suggests that information could flow from the superphysical, through the hyperphysical into the physical (spirit-soul-body). The effect of layered resonance between the planes may be what Burr detected as the life-field, which he measured in the 1930's using valve operated, high impedance volt meters.

We have here a model suggesting a dynamic interplay of forms, a continuum of reciprocal interaction and exchange, via resonance between fields of information on different planes of reality. Information from the hyperphysical could be downloaded via resonance into the physical DNA molecule; conversely information could be uploaded from biological lifeforms back into the hyperphysical. This model has enormous scope for redefining our understanding of the body, the nature of animating intelligence, and the soul.

DNA resonance could shed new light on the *Akashic records,* an idea in Yogic philosophy that all human events, thoughts, intents, words and emotions that have ever occurred – in the past, present, or future – are recorded in a non-physical plane of reality called the *Akashic*. Information from the physical could be uploaded to some form of hyperphysical memory network, equivalent to a computer hard drive, through DNA Resonance.

The information exchanges between planes of energy through resonance might help explain how – as already proposed by Sheldrake as morphic resonance – natural systems, such as termite colonies, or pigeons, orchid plants, or even insulin molecules can inherit a collective memory from all previous things of their kind.[4]

An example of morphic resonance occurred in Southern England. In 1921, in Swaythling, near Southampton, someone reported a milk robbery to the police. A hole had been punched into the cap of a bottle of milk on the victim's doorstep and the top layer of cream had been stolen. The police suspected local youths but the culprit was eventually discovered to be a blue tit. Over the ensuing years sporadic reports of the early morning milk robberies popped up in various parts of the South East of England. Then in the 1950's something extraordinary happened. Milk bottles were being raided by blue tits, not only in England, but throughout Europe. Wherever milk was left in bottles on doorsteps it was liable to have the cream removed through a neat little hole in the cap. In the 1960's blue tits were following milk floats and it was hard to find an unpunctured bottle of milk on the doorstep, unless you got there before the birds.

Research into the learning ability of blue tits indicated they didn't appear to learn this new feeding habit by observing each other. The rapid spread of the new behaviour throughout England and across the channel into Europe seemed to suggest there was some other mediating factor at play other than one blue tit mimicking another blue tit.

4 Sheldrake, Rupert *The presence of the past: Morphic resonance and the habits of nature,* Icon Books, 2011

The proposition, to explain this phenomenon of non-local transmission of behavioural information between members of a species, is that information from a single species individual in the physical domain was uploaded into its hyperphysical brain by DNA resonance. The hyperphysical brain of the species individual, with the newly discovered information, then acted somewhat like a nerve cell in a brain.

The proposal is that the hyperphysical brains of all members of a species, in this case *Cyanistes caeruleus,* the blue tit, are connected in a *hyperphysical species network,* somewhat like the nerve cells in a brain. Then just as one nerve cell in a brain can light up with information received from a sensory cell and pass it on to other nerve cells, so in the hyperphysical species network, information coming from one blue tit, via DNA resonance, could be passed onto other blue tit hyperphysical brains, from where it could be downloaded into their individual physical brains. In this way the learning from one species individual could be passed onto others in a non-physical manner. This could be a *fractal pattern* where, within a species, all the hyperphysical brains of, for example, the blue tits, are networked like nerve cells in a brain.

To reiterate, the hyperphysical brains of blue tits are imagined to be linked together in a hyperphysical network, whereby each individual brain is acting like a single cell in a brain. It is as though the hyperphysical brains of the entire species are networked in the same way as the cells in an individual brain within the species. This is the fractal pattern where a *species brain* could be acting like an individual brain.

The *hyperphysical species networking* could apply, not only to blue tits but to every species of all biological organisms. Hyperphysical species networking could account for the phenomenon that Sheldrake described as *morphic resonance.* This could also account for the *Deva Concept* that all members of an animal species have a single overriding spirit,

Hyperphysical species networking could involve *the hundredth monkey effect,*[5] where information learnt by one species individual isn't immediately passed onto all members of the species. It seems to require a minimum number of individuals to respond to a newly acquired behavioural pattern before the whole species acquires the information.

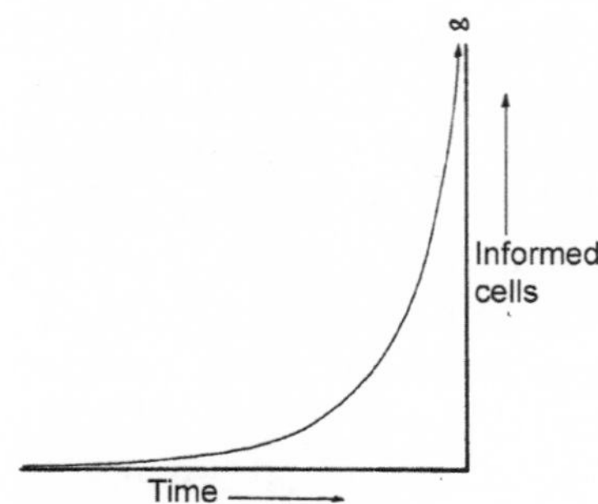

Every individual brain responds to stimulus according to the mathematical law of *exponential growth* as expressed in the *'J' curve*. The first brain cell would inform its immediate neighbours and they then pass the information onto their surrounding cells. Over each interval of time the rate of information transfer accelerates until a *tipping point* is reached. That is when the number of informed cells pass round the curve in the 'J' and the exponential growth occurs.

It makes sense that the passage of information through a *hyperphysical species network* would follow this same pattern of exponential growth. That would explain why it took years for the information to pass from the blue tit population in and around Southampton to other blue tit populations in South East England, until when in the early 1950's, seemingly all of a sudden, the blue tit community of Europe, in entirety, knew how to feed on milk by pecking through milk bottle tops. World War II, in the 1940's might have delayed the exponential curve by reducing the total

5 Keyes K. *The Hundredth Monkey*, Camarillo DeVorss, 1984

number of milk bottles left on doorsteps. This could explain why the tipping point didn't occur until the 1950's when high volume milk deliveries were firmly re-established after the war.

Hyperphysical species networks may be involved in the process of evolution. They may explain how evolutionary changes occur simultaneously in numerous species individuals and why an entire new species can appear intact. DNA resonance may be the mechanism whereby information is uploaded and downloaded between species individuals and the species group, acting as an interconnected whole. The physical plane may be where new opportunities are discovered by individual members of a species and where they are tested by natural selection but the hyperphysical plane may be where the information and experiences of the group are processed and integrated within the species as whole.

Following this line of reasoning, hyperphysical species networks could also be informed and influenced by the super-physical planes of reality. The theory of DNA resonance may help explain how the driving forces of evolution operate between all three levels of reality: the physical, the hyperphysical and the superphysical.

Advancement of a species could occur as follows. Creative innovation and design could happen on the superphysical plane. These could be tested on the physical plane through natural selection. Processing and integration of the tests and trials would then occur on the hyperphysical plane sandwiched in between. The information would be moving in a fluid exchange, back and forth, between the planes via DNA Resonance. The hyperphysical, acting as an Akashic archive, could hold a record of all species events in the hyperphysical species network, thus acting as a species memory.

The superphysical is predominantly plasmic matter; therefore it would operate mainly in terms of *plasmic electric fields*. As previously explained, if these superphysical plasmic fields are alive and endowed with animating intelligence, forming, guiding and modifying templates, to impact the material realm, then they could be compared to the gods as recorded and embellished in mythology. If this were so then perhaps the gods were involved in a process of creative evolution operating through DNA resonance. Is it possible these fields of vibratory intelligence are the source of change in the DNA of cells through resonance? These ideas are conjecture but they do present an argument as to how, through DNA resonance, intelligent 'plasmic field' beings in the superphysical dimensions could play a role in making creative design changes in the evolutionary progress of biological species.

Peter Hewitt proposed that the pastoral god Pan was a metaphor for how evolution is creatively operating through DNA resonance. In *The Vortex: Key to Future Science,*[6] he suggested that if Pan represented an intelligent living plasmic field intervening in the play of evolution, the pipes of Pan could symbolise DNA resonance. Peter said, "When Pan plays the same tune, the species remains the same. When Pan plays a different tune, the species alters."

This thinking may be anathema to materialists; however people who still believe in materialism are like the proverbial frogs in a well that reject the existence of the ocean until experience defies their prejudice. There are progressive writers and thinkers out there, and these include Graham Hancock. Hancock asserts that the conditioning of scientific materialism has influenced 'educated people' to reject the possibility of other realities that are more subtle and multidimensional than the apparent material world. In his foreword to Gregory Sams' *Sun of gOd,*[7] Hancock wrote:

"My personal experiences have opened my eyes to the possibility that the fundamental unit of reality is not matter but the spirit that animates and organizes it, and that there is no dichotomy between spirit and matter because these realms are so thoroughly intertwined and promiscuously interconnected. The problem rather is one of perception, that the logical, positivist, empiricist bias of Western science has so conditioned us to focus on gross matter, as though there is nothing else, that we have become impervious to the subtler fields of spirit that interpenetrate and surround it."

6 Ash D. & Hewitt P. *The Vortex: Key to Future Science,* Gateway Books, 1990
7 Sams G. *Sun of gOd,* Red Wheel Weiser, 2009

Materialistic scientists may object to the ideas proposed herein but one has to bear in mind that their materialistic philosophy has been demolished and the major theories of the 20th Century, endorsed by 'blind-faith-followers' of materialism, lie in ruins. In all honesty they are not in the best position to have their objections to the spiritual and supernatural taken seriously anymore. Like the religion that went before it, science only has a voice if it has credibility. Science, like religion, tells a story and the story of materialism in science may have to give way and allow a new story to come through that lies somewhere between science and religion. This is certainly true of medical science. Whereas physics and biology are of academic interest, medicine impacts the lives of us all.

If DNA resonance can provide an explanation for how cells differentiate into tissues and organs maybe it could help us to better understand disease. Is it possible that a weakening or disruption of the biofield could predispose individuals to disease? By stimulating the biofield through resonance, is it possible we could strengthen or relax the fields within and around the body to help restore and maintain optimum health?

The model of DNA resonance, operating through the human biofield could help provide a scientific understanding of how some modalities in alternative medicine work. This could also help to explain how healing works.

A scientific account for alternative medicine and healing is needed. In 1986, the British Medical Association published a report on alternative medicine,[8] in which they concluded that while the therapies were effective they could not be endorsed because they were unscientific.

The problem is not the alternative medical practice. Alternative medicine is very effective. The problem was and still is the unproven and unscientific materialistic paradigm that has been used inappropriately as the scientific benchmark to determine what is and what is not scientific.

8 BMA Board of Science, *Complementary Medicine: New Approaches to Good Practice*, Oxford University Press 1986

CHAPTER 12
ALTERNATIVE MEDICINE

Alternative medicine is popular and effective, but within the limited framework of the moribund materialistic paradigm it has proved difficult, if not impossible, to explain how alternative therapies work. Acupuncture and homeopathy are outstanding examples. However, if we expand our conceptual horizons to include super-energy, the biofield and DNA resonance, then acupuncture and homeopathy become easier to understand.

While many doctors and scientists dismiss acupuncture and homeopathy as a 'placebo effect,' there is a growing serious interest in these therapies. There use comes under the regulation of Complimentary and Alternative Medicines: CAMs[1],[2],[3] and they are now incorporated into the NHS.

The research of Harold Saxton Burr,[4] may help to shed some light on how acupuncture works. His discovery of a life-field, now more commonly known as the biofield, acting somewhat like an electric field, could be a key because Burr's work suggests the life-field or biofield may be subject to the laws of electricity.

1 *Sixth Report – Complementary and Alternative Medicine;* House of Lords Select Committee on Science and Technology, November 2000

2 Lewith G.T. et al *Complementary Medicine: Evidence base, competence to practice and regulation,* Clin. Med., 2003 May- June (3): 235-40

3 *What Is Complementary and Alternative Medicine?* National Centre for Complementary and Alternative Medicine

4 Burr H.S. *Blueprint for Immortality,* Neville Spearman, 1972

If the same laws of physics apply to super-energy as they do to physical energy and it is only the intrinsic speed of energy, between them that is different, this makes it possible to explain acupuncture in terms of electric field effects. In physics, we are taught that electric fields are evenly distributed over spheres, but if there is a spike on the sphere then a charge will accumulate on the spike. When that occurs, a potential difference builds up between the spike and the body of the sphere. Electrical energy is then able to flow down the *potential* gradient between the spike and the sphere. These simple laws of electrostatics could help in understanding acupuncture.

If a super-energy biofield overshadows a physical body the electrostatic principles would suggest that potential gradients in super-energy could develop between the body and extremities on it such as ears, fingers, toes, nose, and the limbs. And indeed, Burr measured voltages occurring between digits, limbs and the head and body.

Burr's research seemed to indicate a flow of some sort of energy down the potential gradients between the head, the body and the extremities. Could this be super-energy? Is it possible that a biofield resonance occurs between the hyperphysical body and the flow of electric energy in the physical body, which enables a flow of super-energy through the life-field? Could this correspond to the *Chi* life-energy, which the Chinese knew could restore health and vitality when manipulated?

The Chinese discovered flow lines of this life-energy. They realised Chi was flowing between every organ in the body and the limbs terminating in the hands and feet. They also discovered that Chi flowed on the head to the ears and nose. It seems the flow of Chi was following electrostatic principles.

The Chinese called the energy flow lines, or channels of Chi, *meridians*. They found that if they stimulated or sedated the flow of Chi in the meridians they could restore the balance of Chi – the life force energy – in the organs. They did this by inserting needles into specific points on the meridians related to the organs. This system of healing in China came to be known as acupuncture.[5]

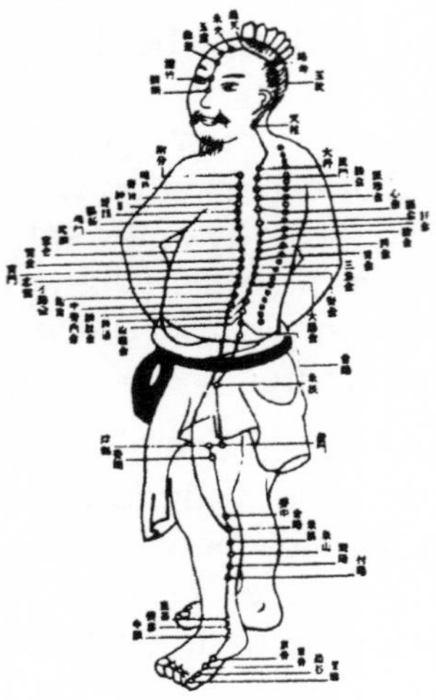

The action of acupuncture and the flow of *Chi* may be supported by the *interstitium,* a network of fluid compartments just under the top layer of the skin which connects to tissue layers lining the organs, gut, blood vessels and muscles. The interconnecting interstitium compartments are supported by a mesh of strong and flexible proteins that are capable of generating electric currents. The interstitium also produces and conducts the flow of *plasma* lymph – capable of conducting electricity – between the skin and other tissues in the body. In acupuncture, the electric activity of the interstitium may be stimulated or sedated to increase or reduce the flow of currents of the life force in it between the skin and major organs of the body.

Another therapy, based on the principles of acupuncture, is called *acupressure.* In acupressure the Chi, or life-energy, is

5 Geng J., *Selecting the Right Acupoints: Handbook on Acupuncture Therapy* New World Press, China 1995

balanced by massaging acupressure points on the skin that are related to specific organs.

Meridians – the life energy flow lines – related to every organ in the body terminate at the feet. Applying acupressure to the terminal point of a meridian on the soles of the feet appears to stimulate healing in its associated organ. This system of life energy balancing is called *reflexology*.

In 1986 when the British Medical Association reported on alternative medicine and said they could not support it because it was 'unscientific', they cited reflexology as a case in point.[6] But millions of people wouldn't pay for reflexology and other alternative therapies if they didn't work.

The problem is that drug orientated orthodox medicine, represented by the BMA, is threatened by alternative medical practice and the medicine the BMA represents is also aligned with the outmoded myth of materialism. A new scientific paradigm is needed to make alternative practices intelligible within a scientific framework. This applies to homeopathy as well as reflexology.

Homeopathic practitioners are accused of fraudulent practice by materialists despite an overwhelming number of people testifying to the efficacy of homeopathy. The new super-energy paradigm can help explain how this remarkable therapy works.

Homeopathy was established 200 years ago by a German Dr. Hahnemann[7] on the principle of 'like cures like'. Hahnemann found if a substance, which caused a particular set of symptoms was greatly diluted, then a diluted form of the substance could also cure the symptoms.

6 BMA Board of Science, *Complementary Medicine: New Approaches to Good Practice*, Oxford University Press 1986

7 Gupta H. and Hahnemann S. *Medicine for the Wise: Hahnemann's Philosophy of Diseases, Medicines and Cures,* Create Space Independent Publishing Platform, 2014

For example, arsenic will cause severe stomach pains, but Hahnemann discovered that an extreme dilution of arsenic could also cure stomach pain. Hahnemann found that the more he diluted his remedies, the more effective they became. This was because he *potentised* his homeopathic remedies by percussing (a procedure of mixing that involves pounding) them in the homeopathic diluting procedure.

Dr Samuel Hahnemann

The way homeopathy works is that in the process of *potentising the medicine*, the super-energy field of the substance (that which is causing the symptoms) is impressed on the super-energy field of the substrate (a neutral material such as sugar or water), in which it is diluted. The energy signature of the substance acts like a 'photographic positive', which is imprinted onto the energy field of the sugar or water substrate, by the pounding or energetic mixing in the potentising process. This produces an 'opposite' energy impression in the substrate, the equivalent of a 'photographic negative'.

With increased dilution, the positive is decreased. With more *potentising* the negative is increased. In this process the concentration of 'symptom causing substance' is reduced while the quantity of opposite 'symptom curing substrate' is increased. This explains why homeopathic remedies become more effective as they are diluted and potentised.

Professor Jacques Benveniste, a medical researcher in France, performed experiments that appeared to validate homeopathy. However, his demonstration in the lab, of the efficacy of homeopathic principles, led to a veritable visit from the Inquisition. John Maddox, editor of Nature, along with an Australian illusionist and leading sceptic, James Randi, descended on Benveniste's laboratory outside Paris. They turned his lab upside down and pulled his records apart. A campaign of disinformation soon followed. Benveniste was discredited. He lost his job and his reputation was ruined. Such is the determination of the scientific establishment, coupled with magicians, to demonise anyone who dares to challenge their pseudo-scientific world view based on the disproven philosophy of materialism.

Dr J Benveniste

The crucial distinction the scientific establishment is missing is that they do not appreciate homeopathic medicines act energetically not chemically. Like many other alternative therapies, homeopathy is concerned with healing by balancing and restoring the energy template of the body and the balance of super-energy in the biofield body template.

Using the super-energy model, other alternative therapies could be accommodated within a scientific framework of understanding. Hands on healing, for example, could be understood in terms of a direct transfer of super-energy from the biofield of a healer to the biofield of a patient. The common denominator being that the biofield can act as a conduit for super-physical plasmic fields existing outside of space and time. These could provide an unlimited reservoir of animating life giving energy, which may be accessed through the various means and modalities of healers and alternative health care practitioners.

CHAPTER 13
HEALING

My father, Michael Ash, was a Harley Street doctor. He was one of the earliest medical practitioners of acupuncture in the UK and he was also a gifted healer.[1] I recall him hovering his hands over a patient, a few centimetres, above the area to be healed. He moved his hands to stimulate a flow of healing energy. He explained to me that the science of healing was similar to moving a magnet in order to generate a current of electricity.

Dr. Michael Ash

It intrigued me that my father could heal people at a distance. I now realise distant healing is possible because super-energy fields exist outside of physical space and time. Beyond the limitations of physical space and time the biofield of a healer, like my father, could join the biofield of his patient by intent. Super-energy could then be transferred from my father's biofield into his patient's biofield.

DNA resonance could facilitate the flow of super-energy into the cells of the patient revitalising them and rejuvenating them and realigning them in the body template. There could be a potentially unlimited reservoir of super-energy that can be accessed by the two biofields connected by intention, while the separation between the healer and the patient would not effect the connection of their super-energy biofields.

1 Ash M., Health, *Radiation and Healing*, Darton, Longman & Todd, 1963.

There are remarkable healing stories in the Gospels where Jesus was able to rejuvenate people instantly. One story of his ability to heal someone at a distance is recorded in Chapter 8 of the Gospel of Matthew. An officer of the Roman Guard approached Jesus and asked him to heal his servant boy who was paralysed and wracked with pain.

Jesus said immediately he would come and heal the lad but the Roman replied that he was not worthy to have Jesus in his home. He said that Jesus had only to say the word and his servant boy would be healed. Jesus was amazed at the man's faith and told him to go home because what he believed had happened. When the officer returned to his home he found the child had completely recovered.

In my teens, I recall my father practicing distant healing on a child in Texas. He practiced distant healing on a number of patients, every morning before breakfast. The child's father said the child would stir and cry every night at about the time my father would think of him and send healing from England. My dad asked the child's father to record the time when the child awoke. Allowing for the time difference between Texas and England they found the child stirred during my father's healing. The child recovered from leukaemia but sadly died from an infection introduced by a blood transfusion.

My father was a close friend of Harry Edwards who was also an exceptionally gifted healer.[2] Harry practiced from his healing centre at Shere in the Surrey Hills. I was introduced to Harry Edwards by my father and my father told Harry I was going to develop a science to explain healing. My father reminded me on a number of occasions that the onus was on me to develop a scientific explanation for healing

In my twenties I was able to explain to my father that in order to understand healing we had to embrace the idea of energy as pure motion. After reading the *Tao of Physics* I was able to tell

2 Edwards H., *Harry Edwards: Thirty Years a Spiritual Healer*, Jenkins 1968

him that Buddha had understood this concept millennia ago. Quoting Fritjof Capra from the *Tao of Physics:*[3]

"Buddha formulated a philosophy of change. He reduced substances, souls, monads and things to forces, movements, sequences and processes, adopting a dynamic conception of reality."

Buddha also said, "Right understanding is part of enlightenment." In that spirit I began to fulfil the task my father set me by using ancient Indian philosophy to explain how matter is formed of energy. Having studied Yogic philosophy,[4] I understood the smallest particles of matter were fundamentally vortices of energy. I realised that subatomic particles were pure movement spinning in a vortex, despite the illusion of solid matter.

As I developed my physics and understanding that particles of energy were particles of pure movement, I came to realise that the speed of light we measure must be the speed of energy in matter and light that make up our world. That, I explained to my dad, was why Einstein asserted mass, space and time are all relative to the speed of light. I then reasoned that speeds of movement, faster than the speed of light, could set up worlds of super-energy and that these higher levels of super-energy could provide an account for healing and other spiritual and supernatural phenomena.

I told my father that the spiritual energy flowing through him, and healers like him, could be coming from worlds of super-energy; worlds beyond the speed of light. I told him how, because the speed of light is slower than the speed of super-energy in the spiritual worlds, our world would be part of the spiritual worlds, and spiritual beings in these worlds would see us even if we couldn't see them.

3 Capra F. *The Tao of Physics,* Wildwood House, 1975

4 RamacharakaYogi, *An Advanced Course in Yogi Philosophy,* Fowler 1904 (Cosimao Classics Philosophy 2005)

My father, who was in the process of appointment as a physician to King George VI (which was never completed due to the premature death of the King), enjoyed mixing with psychics and spiritualists as much as he enjoyed mingling with medical practitioners. He also had a profound interest in physics relating to radioactivity and geopathology. He was the first doctor to raise alarm about the danger of living above uranium in the West Country Cornwall. He was also the first medical doctor to alert to the danger, in the early 1950's that cigarette smoking causes lung cancer.

My dad introduced me to nuclear physics when I was six. He was a polymath; as well as being a physician and physicist he was also a geologist and a renowned healer. My father also taught me how to heal so there was no separation in my mind between science and healing. I perceived healing in terms of the operation of forces. Working with my hands, as a healer, in the way my father taught me, I could feel the force he worked with as a tingling in my fingers, as did my brothers and sisters who he also taught to heal.

My grandson, Ezra, has inherited his great grandfather's gift of healing. He can also feel 'the energies' in rocks and on the land and he can track nature spirits by sensing their energies with his hands, just as my father did. Ezra says he feels the energies as a tingling sensation in his hand and his fingers.

I can explain the physics of healing to my grandson's generation. I explain that according to the principles of the vortex physics we have a super-energy body as well as a physical body. The super-energy body incorporates an electrodynamic biofield, which acts as a template for the physical body. The principle of simultaneous existence clarifies how the super-energy field can interpenetrate the physical body, while the principle of energy speed subsets suggests that energy will flow from the super-physical into the physical – rather like water flowing downhill or electricity flowing down a potential gradient.

Spiritual healing can occur through resonance. Resonance works through energy exchange between corresponding forms or shapes. According to the new super-energy paradigm, we have a hyperphysical body that acts as an 'identical shape' intermediary, which allows for resonance to occur between the superphysical and physical. This is possible because the hyperphysical body replicates the physical body cell for cell and molecule for molecule but it is based on a higher intrinsic speed of energy than the speed of light that underpins the physical body.

As the DNA in each physical cell is overlaid by DNA in hyperphysical cells, information in the superphysical biofield transmitted to the hyperphysical intermediary body, can through DNA resonance be transferred to the physical body. The vibration in the hyperphysical DNA passes to the physical DNA, not by direct contact but by resonance. This occurs in the same way that a vibration from the tuning fork can pass to the guitar string, tuned to the same frequency, even though there is no direct contact between them.

The hyperphysical body is not affected by aging or disease, or wear and tear incurred by the physical body. It maintains its integrity because it is not in physical space or time and therefore is not subject to physical influences. The hyperphysical body supports the integrity, form and function of the physical body through DNA resonance. The physical body is continually sustained by the hyperphysical body, but with age and stress, disease and injury, the physical body can degenerate beyond the capabilities of the hyperphysical to uphold it. This is where the intervention of a healer can be of benefit.

A 'spiritual' healer can channel superphysical energy through his or her hyperphysical biofield into the biofield of the patient, to stimulate the innate healing ability of the patient's body. Gifted spiritual healers like my father and Harry Edwards in recent times, or Jesus Christ and his contemporary, Apollonius of Tyana, in ancient times, all channelled super-physical energy through their hyperphysical 'soul' bodies into the super-energy biofields of the afflicted people. In essence, healers stimulate the hyperphysical body template to assist it in its innate ability to regenerate the physical body; all of which occurs through DNA resonance.

In the new super-energy paradigm, the biofield, as a resonant expression of the life-force in super-energy, could overlay and transmit healing to the physical body. This powerful idea could account for the aura perceived by many healers and psychics as an energy field surrounding the body.

Psychics and healers sometimes see the aura as a light surrounding the body. An exceptionally bright aura of light, sometimes seen around the head, corresponds to a halo.

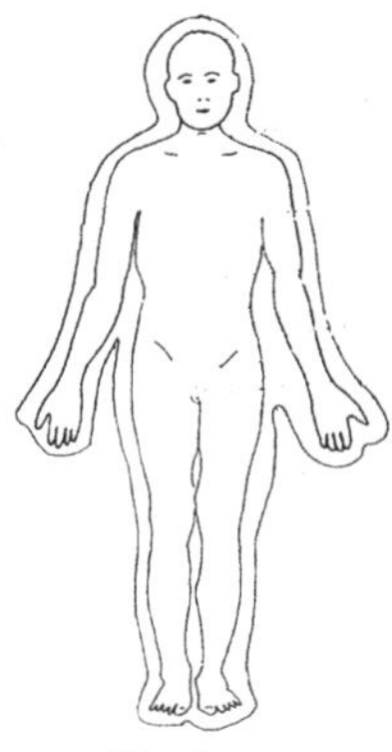

The Aura

Perception of the aura, through seeing or feeling by those with the faculty for psychic perception has occurred since ancient times. Perception of the aura amongst the psychic community could correspond to the Platonic assertion of a *psyche* body template that carries the life force into a body at conception and departs it at death.

Nature spirits, in the form of elementals, constitute the auras or biofields associated with plants. Even non-living things like rocks and pools, places and human constructions can have hyperphysical energy fields perceived as auras. Inanimate objects like a rock or a stool can have an aura with a degree of innate intelligence. This is why some people keep rocks and pebbles as pets. The innate intelligence within hyperphysical auras, associated with trees and plants, can be pronounced which is why sensitive people like H.R.H. the Prince of Wales speaks to his plants and has been known to hug his trees. He sets a good example because respecting and communicating with the plant kingdom can help us all become more sensitive to our environment and its fragile ecology. We can develop sensitivity, through plant auras, to the origin of life and begin to feel its awesome presence everywhere and in everything.

Psychics and healers sometimes perceive colour in the human aura. These relate to emotions and the health and overall disposition of the patient. These colour frequencies in the aura can be used by healers in diagnosing disease. Healers are also aware of broken auras in patients that are ill or are mentally or emotionally disturbed. All too often healers perceive aura damage in association with addiction to drugs and alcohol.

Having the properties of an electric field, the aura can be photographed using equipment that involves high voltage and high frequency electric fields. These resonate with the biofield to generate images of a corona of coloured lights. The difference between coronas generated around inanimate and animate objects is usually registered in the vivid colours.

Aura images were first captured on a photographic plate in 1939 by a Russian electrical engineer, Semyon Kirlian and his wife, Valentina. The electro-photography they developed to capture images of the aura came to be known as Kirlian photography. The vivid colours in the aura are depicted in *aura photography*, which is now widely available through the development of Kirlian electro-photography. In essence Kirlian photography, and the aura photography developed from it, is based on a simple technology for picking up the resonance between a physical electric field and the bioelectric fields of the aura.

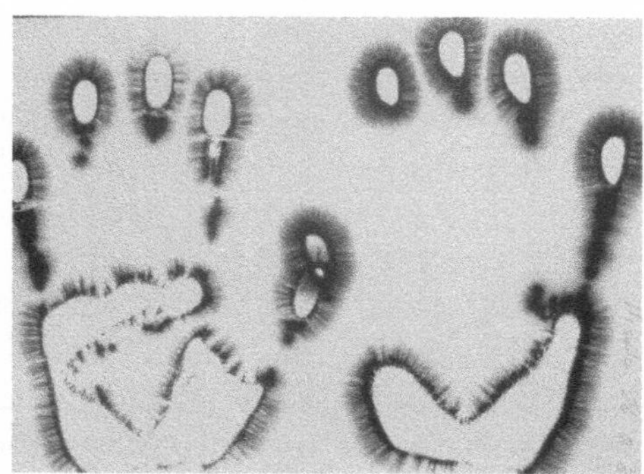

The Kirlians received a Russian patent in 1949 and first published their work in 1958 but because of the Cold War it was

1970 before their work reached the West. While Kirlian photography was endorsed by the Russian Academy of Science in 1966,[5] in the West it was treated with disdain and eventually damned by scientists and sceptics as a hoax.[6]

A professor at the University of California at Los Angeles (UCLA), Thelma Moss, opened a scientific laboratory to research Kirlian photography in the early 1970's. The lab was unfunded and run by volunteers but its line of scientific research into the resonance effects between living organisms and high frequency, high voltage electric fields, was deemed to be outside the bounds of legitimate science and therefore was not tolerated by UCLA. The lab was subsequently closed down by the university in 1979.[7]

To any reasonable individual Kirlian photography would qualify as legitimate scientific research. However, because it challenges the tenets of materialism and is in danger of providing evidence in support of the supernatural it is damned. This is why it is necessary to expose the cracks in physics and challenge the validity of the materialistic science underpinning university establishments in the Western World.

The anti-soul, anti-spirit bias amongst university professors, determined to uphold the prevailing false philosophy of materialism, is filtering down into the levels of compulsory school education. Secondary education should come under review to ensure that parents are not being enforced by law to expose their children to the conditioning of a philosophy, which has in recent times been invalidated by science. Scientific materialism stands on a par with religion, in so far that belief in the doctrine of materialism is based on blind faith rather than experimental facts. Scientific materialism should come under the same restrictions as religious indoctrination in schools and other educational establishments.

5 Juravlev A. E. *Living Luminescence and Kirlian Effect,* USSR Academy of Science 1966
6 Stein G. *Encyclopedia of Hoaxes* Gale Group 1993
7 Greene S. *UCLA lab researched parapsychology in the '70s, A closer Look.* UCLA Daily Bruin Oct.27 2010

BOOK V

SOUL AND SPIRIT

With regard to their (Druid) actual course of studies, the main object of all education is, in their opinion, to imbue their scholars with a firm belief in the indestructibility of the human soul, which according to their belief, merely passes at death from one tenement to another; for by such doctrine alone, they say, which robs death of all its terrors, can the highest form of human courage be developed.

Julius Caesar

CHAPTER 1
ANCESTRAL SPIRITS

Many ancient cultures have left clues for an understanding science that is less materialistic and more congruent with quantum thinking. Australia's indigenous people have an ancient living culture that spans over 50,000 years and their knowledge is mystical and spiritual. It connects people with the land through art, dance, music, secret stories and ritual journeys into the mysteries known as the *Dreamtime*.

In their story of the Dreamtime, the indigenous peoples of Australia anticipated quantum consciousness by appreciating the world more as a dream than a concrete reality. In our modern age we are familiar with virtual reality. The worlds we engage with on computers are just electric impulses. Conscious of the world, as we are through nervous electric impulses, how can we be sure the world we live in is not a virtual reality?

If particles of energy are more like thoughts than things, if they are particle expressions of quantum consciousness – more acts of consciousness than acts of material things – then all reality would be virtual. If that were so there could be many levels of virtual reality in the Universe.

When we wake up from sleep we often recall dreams, which are as real to us in the first moments of waking as is the waking state we arise into. As our consciousness moves between sleeping and waking it feels as though we are moving between levels of reality. If the Universe is more a mind than a material machine then the world we inhabit in the waking state could be just another level of dreaming. We could be living in a dream.

The dreamlike nature of reality is recognised by many quantum physicists. To quote Carlo Rovelli writing about quantum reality in *Reality is not What it Seems*:

"... And yet, instead, it is a glance towards reality. Or better, a glimpse of reality, a little less veiled than our blurred and banal everyday view of it. A reality which seems to be made of the same stuff our dreams are made of, but which is nevertheless more real than our clouded daily dreaming."[1]

Have you ever had the experience of dreaming, the dream vanishes as you wake up but then you slip back into sleep and find yourself back in the same dream again. When we awaken into our daily echelon of consciousness, dreams vanish but perhaps they reappear when our consciousness returns to the realm where dreams normally reside. Do they continue to exist when we are awake or do they vanish out of existence altogether? According to quantum mechanics we can never be quite sure.

Quantum theory suggests that what we take to be real in the physical world could be behaving like dreams at a quantum level. Werner Heisenberg developed quantum mechanics on the basis that subatomic particles may exist only when, like a dream, we are conscious of them. Heisenberg proposed that subatomic particles can only be treated as existing when they interact with something. Otherwise they can, like dreams, be treated as non-existent.

Heisenberg initiated a system of matrix mathematics based on predicting the position of an electron not at every given moment but only in the moment when it interacts with something else. The quantum mechanics he devised was based on the *relational* existence of things rather than their *absolute* existence.

I like to think of Heisenberg's math as the *mathematics of dreams* to the extent that dreams only exist when we are conscious of dreaming; when we actually experience them. Otherwise, to us, the dreams do not exist.

The brilliant – possibly autistic – mathematician Paul Dirac turned Heisenberg's intuitive calculations and matrices of numbers into the formidable formulae of quantum mechanics. These have worked, with incredible success, in practically every

1 Rovelli C. *Reality is not what it Seem*s Penguin 2017

sphere of science, engineering and technology. The computer I am writing on depends on them.

The undeniable success of quantum mechanics attests to the dreamlike nature of reality. The way quantum particles behave does suggest that, at the quantum level, the world we inhabit, in the awakened state, is not unlike the world we visit when we slip into sleep.

Quantum mechanics supports the mythology of the indigenous Australians. They believe we live in a level of the dreamtime and move in and out of dream states between life and death as between sleeping and waking. They went further than common dreaming. In their dancing and their ceremonies they went into trances, and in their altered state they claim to have entered other dreamtimes consciously.

Many people in ancient and modern times believe that in trance and drug induced states that they experience other realities. Is it possible that altered brain chemistry enables the psyche to move into other levels of the Universe? This is feasible in terms of the metaphysical implications of quantum theory. Quantum states occur in the microcosm of the atom. Why not in the macrocosm of the Universe?

Those who still adhere to the pre-quantum materialistic world view account for consciousness in terms of brain biochemistry. Those who are at the cutting edge of quantum thinking are open to the possibility that consciousness exists as a Universal phenomenon and is aware of itself through the mediation of brain biochemistry in animals like ourselves.

When brain activity is modified by sleep, trance or psychoactive substances, consciousness may be temporarily loosened from the constraints of the brain. Awareness of universal levels beyond the physical world might then occur. This presupposes the existence of other worlds of energy beyond our perceived reality. This proposition is a core spiritual belief. The other core spiritual belief is that these other worlds are occupied by supernatural beings much like us.

Australia's indigenous peoples believed that ancestral spirits came to earth and created all things. This belief is common to other ancient cultures, including the Hebrews.

There was a time when the God of the Hebrew Bible was seen not as one God but as many gods. Maybe the gods of the Bible were beings from another dreamtime.

In the original Hebrew texts, the book of Genesis began with the words, "In the beginning elohim created the heavens and the Earth". The Hebrew word *elohim* translates into 'gods' not 'God'. If God is defined as spirit, elohim would be spirits.

God is defined as *spirit* and the plurality of 'God' is clear from follow on verses in the Hebrew book of Genesis where the gods say, "Let us create man in our own image, according to our own likeness." (Genesis 1:26)… "Here the man has become like one of us." (Genesis 3:22)…"Come now let us go down there." (Genesis 11:7). A verse in the Bible where 'God' is described as the 'Ancient of Days' (Daniel 7:9) suggests the spirits we call God are ancient.

The Aboriginal concept of ancestral spirits is that the gods are our ancestors. To the indigenous Australians the creator gods are spirits that are linked to us and our past. They are not separate from us but are a part of us because they relate to our personal past. As Mark Blaisse put it, "They are part of our history, our origins and our fundamental being. They are us."

Imagine our world as a three dimensional virtual reality created in an enormous universal computer of some sort so that everything appears as real as it is to us yet is not real in an absolute sense. Imagine this virtual reality was set up by a group of technologically advanced beings who then decided to enter the virtual world they created.

Now imagine we are them, here and now, experiencing the virtual world in the human biological machines we created in our distant past. The creator gods that set up the virtual reality, the dream state we are in now could be us. If this were so they would be our ancestors and represent our past.

In this potent creation myth the creator gods would not be apart from us, they would be a part of us. If this mythology were true then in our ancestry we would have created this world for our own purpose. Were this so then we would be entirely responsible for everything that happens in this world and for everything that happens to us. We wouldn't be able to blame

God, the devil or anyone else as they would all be aspects of our own being.

Our world could be likened to the TV programme, *Game of Thrones*. In *Game of Thrones* a group of people play in front of cameras that relay images of them for broadcast. These are replicated as impulses in millions of television sets. TVs are full of real life drama but none of it is real. Maybe human history is like a television drama. After all we are just energy; nothing but a bunch of impulses!

Did we, in our ancient ancestry, create life on earth? Is our history just for our entertainment or did we intend it for our education? Hindus view God as creator, preserver and destroyer. We all have that in us! Is human life on Earth purposeless, or does it have a purpose?

Maybe we used creative evolution to develop human bodies as tickets to theatre Earth. Maybe we are the playwrights and the producers, the directors and the actors in the drama of human history. Maybe we are also the audience. For some of us this maybe just one of the ways we keep ourselves entertained in eternity. For others it may be a priceless opportunity for advancement. If we are spirits having a human experience maybe we are creating the 'divine' purpose of our own lives. If we are divinity evolving through humanity perhaps at an ultimate level, we are the origin of life, the origin of our own lives and the origin of what we perceive to be God.

In *The Origin of God*[2] Laurence Gardner points out that 'el' translates as: *Shining One* and 'elohim': *The Shining Ones*. According to Gardner, in the Hebrew Bible 'God' addressed himself to Abraham no less than 48 times as *El Shaddai*, which renders as *The Shining Lord of the Mountain*. In Christian Bibles this appellation has been shortened simply to *The Lord*. Appearing as he did to Abraham, as a man, El Shaddai may have been an ancestral human descending not from a mountain but from a higher dimension of reality.

2 Gardner L. *The Origin of God*, Dash House, 2010

The word *el* in Hebrew also signifies 'angel'. That is why the primary archangels were called, Michael, Gabriel, Raphael, and Ariel.

Verses in the Old Testament of the Christian Bible make it clear that 'The Lord' who spoke to Abraham was an angel:

"Lay down the knife; don't hurt the lad in any way" the **Angel** said, "for I know that God is first in your life – you have not withheld even your beloved son from me."

Genesis 22: 12

Then the **Angel** of God called again to Abraham from heaven. "I, the Lord, have sworn by myself that because you have obeyed me and have not withheld even your beloved son from me..."

Genesis 22: 15-16

These early verses in the Bible make a clear distinction between the God, who Abraham held first in his life, and the Lord who spoke to him. These verses make it clear, to anyone who cares to study the Bible with due diligence, that the Lord in the Bible is not God but an angelic messenger of God. The Bible God defined as 'light', 'life' and 'spirit' cannot be the source of Universe of Energy because light, life and spirit are forms of energy. It is more likely the Bible God is an angel or a group of angels. If they made us then the Bible God could be our 'angelic parentage'. This is borne out in the New Testament where Jesus refers to God as his 'Heavenly Father'. Is it possible that we are descended from angels? Can it be construed from the Bible that angels are our heavenly progenitors? Are we proto-angels, baby angels if you like, that have been seeded onto the Earth to gestate and mature as angels until eventually, after many lifetimes of learning, we take our place in the celestial choirs? Do humans view their angelic parents as gods much as children view their terrestrial parents as gods? Perhaps it is time for us to consider the possibility that we are gods in the making, and maybe even that we are very nearly made.

CHAPTER 2
ANGELS AND FAIRIES

The concept of spirits can be accommodated in the new paradigm of super-energy, as beings formed of energy based on speeds of motion faster than the speed of light. This fits with the traditional mythology of spirits in which the dimension of separation between them and us was considered to be a factor of speed. In the ancient religious lore of spirits it was said that: Angels are all around us but we don't see them because they are moving too fast. In the case of fairy folk lore it was said: Fairies live at the bottom of the garden, but we don't see them because we are too slow and clumsy.

Everything is a myth, your body, your home and family, your possessions and everything you hold dear is a confabulation. Everything is formed of energy and energy is more akin to the nature of myth than material. So when it comes to myths it is not a matter of whether they are true or false so much as whether we think they are worth believing in. In *Sapiens: A Brief History of Mankind,*[1] Yuval Noah Harari, pointed out that money is a myth. Nonetheless, because we need money, and have to work hard for it in order to survive, all of us – apart from the likes of Jesus, Buddha, Francis of Assisi and Daniel Suelo[2] – believe in that particular myth.

Dismissing angels and fairies doesn't serve anyone. Even as myths they are delightful and bring joy and comfort to millions if not billions of people throughout the world. My attempt is not to prove that angels and fairies are real but rather to create a climate in science that supports those who believe they exist; people like

1 Harari Y. *Sapiens: A Brief History of Mankind,* Vintage 2014

2 Sundeen M. *The Man who Quit Money* Riverhead 2012

William Bloom. In *Working with Angels Fairies and Nature Spirits*[3] Bloom writes:

"The angel world does exist. It is part of the fabric of nature and the universe. It is part of the creative beauty and juice of all life. Knowing this is deeply relevant to the social and environmental problems of our time. It is also relevant to the dynamic need of every aspect of life to fulfil itself. And there is no need for any of us to be embarrassed because we understand and work with the angelic realities."

I believe in spirits, including angels and fairies, because I can provide a reasonable account for them in my story of super-energy, grounded in my vortex physics. I am also impressed by the fact that my account for them, as living, intelligent beings based on faster energy speeds, appears to be supported in ancient folklore. I am also humble enough to consider it possible that some people who claim to see angels and fairies may not be peculiar or prone to imagine things; they may be having a genuine experience. I delight in the idea of having a physics of the fairies. Finally I had an experience, which suggested to me that fairies do exist. It happened not when I was daydreaming or going off the rails but when I applied myself to cleaning up the environment.

Most people don't see the spirits and therefore disbelieve in them. Sceptics would say those who see spirits and hear voices are either deranged or deluded, or are just imagining things. Others believe in spiritual beings more as an escapist, romantic illusion than a practical reality. Both those opinion positions are irksome to those who take fairies and angels seriously. In the words of William Bloom:

"I am also tired of writings and conversations about angels that have no awareness of environmental, social and psychological realities. The angelic realm is very beautiful, but it is a beauty that needs to inspire us in the real world rather than lift us into escapist illusion."

3 Bloom W. *Working with Angels, Fairies and Nature Spirits* Piatkus 1998

The question for most of us who do not see spirits is are the people who do see spirits imagining them or do they actually see something we do not see? Is it possible that some people have extraordinary powers of perception?

In *The Doors of Perception*[4] Aldous Huxley described the brain as a filtering device which prevents us perceiving other realities apart from the physical. Could it be that some people are wired up differently to others? Maybe the chemical filters in their brains, that are supposed to block out other levels of reality, for our day to day survival, are 'lifted'. Is it possible we don't all perceive the same reality? Maybe the difference between psychics and sceptics could be something to do with their brain chemistry.

Different people may have different ways of viewing the world. It could be that most people are conscious only of the world of physical energy, whereas a minority may be conscious simultaneously of the world of physical-energy and super-energy. Psychic faculties are more often than not inherited. That suggests psychics may have inherited a different type of brain chemistry to most of us. Rather than being an aberration, the possession of psychic faculties may represent a type of brain that is more receptive to other planes of reality than the brains of most people. In times gone by, people with psychic faculties were thought to be 'gifted'.

No one knows for sure the cause of psychic awareness. It could be that worlds of energy exist beyond the speed of light which a few see but most don't. It can be hard for the minority who see into other worlds because they experience discrimination from the majority of people that don't have 'the sight'.

Is it possible that angels and other spirits, like elementals associated with plants, rocks and places, actually exist as living forms of super-energy? Do some people with heightened awareness actually see or hear them? Rather than moving too

4 Huxley A. *The Doors of Perception* Penguin Books 1959

fast for us to perceive, is it possible that the intrinsic speed of their energy is just too fast to be contained in our physical continuum of space and time? Is it possible angels and nature spirits are not people with wings but are advanced forms of intelligence that can manipulate the brains of humans so that they are recognisable by winged form?

The purpose of this book is not to attempt to convince those who do not see and don't want to believe. Its purpose is, rather, to construct a rational frame of understanding, based on the vortex theory, so those who do see otherworldly things do not think they are going mad.

It is not necessary to actually see spirits in order to be aware of them. I don't see spirits, such as the spirits of nature like fairies, but I do feel them. I sometimes go into a trashed space, like a wood strewn with litter, and I clean it up. I feel the depression when I first enter a wood where people have fly-tipped or littered. When I clean it up I feel the atmosphere lift. Maybe you have had a similar experience to me?

Although one of my daughters saw a fairy when she was little and I have a sister-in-law who saw lots of fairies in Cornwall and one of my brothers and his wife saw fairies in St Nectan's Glen in Cornwall, I never saw a fairy. As a scientist and a skeptic I would feel honour bound to question my experience if I did. However, in the summer of 1992, I had an experience I have found very hard to explain away.

One sunny day in July 1992 I parked my campervan on the double yellow lines just below Glastonbury Tor, switched off the engine, locked up and began to climb the Tor. As I did so I noticed litter was strewn about everywhere. I felt depressed and sad about the dereliction of that sacred site so I decided to go back to my van and collect a black bin bag and dedicate the rest of the afternoon to cleaning up the Tor.

As I picked up the litter I amused myself by chattering to imaginary fairies. I told them, as if they were real, that I was clearing the Tor for them. The more I chattered the happier I became but I was careful to stop laughing and talking to the fairies when anyone passed within earshot because I didn't want people to think I was stark raving bonkers.

I didn't see any fairies. I had no reason to believe in them. I was just having fun talking to them as though they were real. I have more faith in *feeling-is-believing* than in *seeing-is-believing*. I felt extraordinarily happy the day I allowed myself to go off with the fairies as I cleared up Glastonbury Tor.

Over several hours I collected a black bag full of assorted litter from sweet wrappings and crisp bags to drink cans and bottles. I tied up the bag and left it on the roadside for the Council to collect. I felt I had worked for my parking so I decided to spend the night on the double yellow lines.

It was after ten and I was about to go to bed when there was a knock on the door. My immediate reaction was it must be the police coming to move me on but I was mistaken.

When I opened the door there was a long haired hippie staring up at me, "I am up from the town," he said, "The fairies asked me to come and thank you for cleaning the Tor."

He didn't stop to say anything else but turned round and disappeared into the night. Even if he had seen me on the Tor and heard me chatting to the fairies, how could he have known, hours later, that I was in that campervan?

In ancient times most people believed in the spirits of nature. There was never a divide between the natural and supernatural. Only since the advent of science and Bible based religion has this division occurred.

Religion, as well as materialism, has driven a wedge between natural science and spirituality. In his foreword to *The Origin of God*,[5] Laurence Gardner wrote:

"In erstwhile Mesopotamian, Canaanite and Egyptian thought, the inexplicable divine was always thought to be manifest within Nature, and Nature enveloped both the gods and society. But this belief was shattered for all time by the priests and biblical compilers, who forsook harmony in favour of subservience. Hence, the balance of relationship between

5 Gardner L. *The Origin of God* Dash House 2010

humankind and the phenomenal world was destroyed and what was ultimately lost was integrity."

I consider that thoughts can act like viruses. Systems of belief can infect people. I think we should take care to protect ourselves and our children from becoming infected with the dubious mental viruses of monotheism and materialism that have become so pervasive in our society.

My objective is not to create a new mind virus but to counter the established mind viruses with the equivalent of a vaccination; something that has elements of both of them. Accept what I say only to the extent that it offers an opportunity to reconsider the spiritual and supernatural in a scientific context; as a way to embrace ancient knowledge rather than dismiss it as irrelevant. Feel into what I say. Exercise discernment. This can be achieved through a better understanding of the Universe inside us as well as outside of us because the big and the small and the inside and the out are all connected in the single whole. In science this is described by the principle of *fractals*.

CHAPTER 3
THE FRACTAL UNIVERSE

The description of angels acting as messengers of God conveys a prodigious idea that spirits, as plasmic lifeforms, could be living electric transmitters of universal information. Angels, as spirits, could be a representation of the branching of Universal Consciousness by means of fractals.

The hermetic principle, as it is above so it is below, as it is below so it is above, enunciates the fractal principle; the principle of infinite regress. It is also the principle of the whole being represented in the parts. The fractal principle is central to Chaos theory and is illustrated by the Mandelbrot set. [1]

In 1979, Benoit Mandelbrot[1] fed the answer to an equation back into the equation to produce an endlessly repeating pattern on a computer. The now famous pattern, named after him, is edged with minute projections, each of which can be expanded into a repeat of the original pattern. The fractal pattern can repeat itself endlessly.

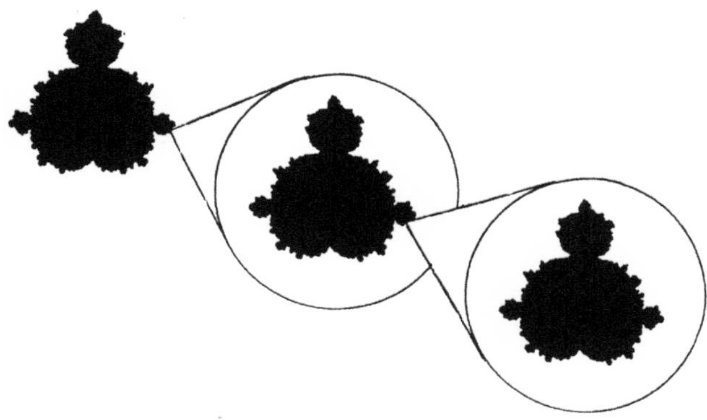

1 Mandelbrot B., *Fractals and Chaos: The Mandelbrot Set and Beyond*, Springer, 2004

The religious description of God surrounded by archangels that are surrounded in turn by legions of angels could be an attempt to describe the Universe in terms of fractals. One way to understand the fractal nature of the Universe, in a religious context, would be to imagine God as a power station, archangels as the national grid and angels as the domestic supply that brings electricity to our homes. Another useful analogy is an optic fibre Xmas tree. Imagine God as the central light and the angels as the optic fibres that distribute the light out to the many points on the tree.

Each human being could be a fractal in the Universe acting somewhat like a Mandelbrot set. Each of us, as a thinking, imaginative being, endowed with the powers of creation, preservation and destruction, may be a scaled down version of what we understand to be God. If this were so, to know God we would need to know ourselves and instead of looking for God outside of ourselves, maybe we should look for God within.

The Egyptian philosopher Hermes Trismigestus is attributed with elucidating the fractal principle as above, so below. He is also attributed with the word human. He described us as hu-man beings; the word 'hu' coming from one of the Egyptian words for 'God' or the 'Divine'. Fractals can help us understand the teaching of Hermes that we are sentient beings with attributes of the Divine; especially the ability to love, act with intent and to think and reason.

The fractal principle could help to explain how we may all be the same one being in many bodies, experiencing the illusion of division and separation. An illusion of separation may be set up by our brain, acting as a filter, to enable each of us to be a unique individual with a free will.

The Mandelbrot set can help us to appreciate how we can be individual and yet still be part of the whole. This universal fractal principle, known to Hermes Trismigestus, was also understood by Gutama Buddha in the third century BCE.

Buddha encapsulated the fractal principle in his Lotus Sutra: "Our existence is identical to the Universe as a whole, and the Universe as a whole is identical to our existence."

So important is this mystic law of life – We are in the Universe and the Universe is in us – that Nichiren Daishonin, in thirteenth century Japan, encouraged people to chant **Nam-myoho -renge-kyo**, which translates as: *Remember the mystic law of life revealed in the Lotus Sutra*. Today millions of people throughout the world, chant this affirmation, which acts to bring unity and harmony to humanity as a whole and peace and well-being to relationships, family and individual human life.

We are so conditioned by the material paradigm to think that we are nothing but a body driven by a brain that we cut ourselves off from the Universe at large. Once we appreciate that each and every one of us, is an image or a fractal of the Universe, we can rise above our limiting materialistic conditioning to know who we really are. From there we can come into an understanding of the soul and the potential for a continuity of life after physical death.

CHAPTER 4
KNOW THYSELF

Socrates is renowned for the maxim "Know thyself." If we are fractals of the Universal Consciousness and are the one being in the many bodies, we could come to know the Universe by getting to know ourselves. This possibility was established by Yogis. In the pursuit of knowledge they employed introspection to glimpse into the quantum world. That was how they discovered the vortex of energy.

In the West the emphasis for gathering knowledge has been placed on exploring the external world. In India, the emphasis was placed on exploring the inner world. In the tradition of Yoga the key to the Universe was discovered, not through high energy physics but through a contemplative process of inner exploration. And the Yogic discovery of the subatomic vortex, through the use of siddhi powers, predates many discoveries in modern physics. Now is the time to bring the scientific approaches of the East and the West together to enable us to come to a greater degree of understanding of the mysteries of the Universe.

In a BBC programme on the Universe (December 2016), Brian Cox said, "The Universe is getting to know itself through us." It could also be said, "We can get to know the Universe through knowledge of ourselves."

In the world today, as the East and West continue to merge, mysticism and science are coming together to offer a more complete understanding of the Universe. A new generation is recognising and adopting the Eastern approach to self knowledge. As they do so they appreciate experientially that the Yogic way of going deep within, in a journey of personal inner exploration, is as important as the Western scientific approach of looking out into the external world.

Through the practice of meditation it became clear to me that there is a distinction between thought and the awareness of thought. Meditation begins when thinking stops. In meditation we can come to realise that at the core of our being we are not thought but the awareness of thought. As we go about our daily affairs we can *witness ourselves.* Rather than identifying with our thoughts, reactions and emotions, we can witness them with impartiality. This is the path to *self-mastery*. That is how we can come to appreciate ourselves as embodiments of 'Universal Consciousness'.

When we come to know ourselves, not as a bunch of thoughts but as the awareness of thought, we can separate ourselves from our thinking. We can benefit from teachers in the Yogic tradition who can help us learn to ignore extraneous thoughts and achieve peace of mind. I recommend the renowned teacher of self-knowledge, Prem Rawat who taught me that I am not my thoughts and showed me how to attain lasting inner peace.

The 17th century, French philosopher, René Descartes, famously said, "I think therefore I am." Descartes epitomised Western philosophy in which thought is considered to be paramount. An Indian Yogi would be more likely to say "I AM therefore I think". In meditation Yogis become aware of their innate being as the 'I Am That I Am' state of conscious awareness. They realise the state of being consciously aware is more fundamental than thinking.

Most of us are attached to our thinking. We confuse our I Am presence with our ego thinking self. This is because, in materialistic science, thoughts are considered to occur only in the brain. According to Yogic philosophy thought is seen to be a universal energy that is finer or faster than light. Yogis also recognise we can receive and transmit thoughts.

Do our brains act like radio sets? Is telepathy a reality? Is it possible some of our thoughts are generated by our brains while other thoughts are received by our brains, which we then identify with as our own? If this were so, where would the extraneous thoughts be coming from?

CHAPTER 5
BODY AND SOUL

If the Universe is a mind it would not be unreasonable to suggest we may be receiving thoughts from sources beyond our own brains. If the Universe is populated by living beings in higher dimensions of reality they could be transmitting thoughts that we receive. Maybe we are individualisations of the universal mind, which survive physical death. Maybe the mind in our heads is comparable to a website embedded within a universal Internet, which is accessed through the hardware of our brain, just as the Internet is accessed by a computer or other interface device.

Perhaps our physical body is like a computer or mobile we trash and replace every few years and our mind is like the website we continue to access with the new computer or smart phone, as we did with the old. Maybe we have many lifetimes and each lifetime writes a new page on our own cosmic site. Is it possible we are influenced in our thoughts by our pages from previous lifetimes? Plato spoke of the psyche as a template for the physical body that comes into the world at birth and departs it at death, richer for the experience of a lifetime. In India many people believe in reincarnation, which fits with Plato's hypothesis.

In the West the individualised mind or *psyche* is sometimes called the *soul.* The suggestion that we have a soul will not appeal to everyone. Many people heap scorn on the suggestion we have an immortal soul. There is peer pressure in schools, colleges and universities to maintain the anti-soul consensus view of mainstream, materialistic science because most scientists, academics and philosophers are committed to materialism. This is despite the fact that the philosophy of materialism is no longer supported by science. The anti-spirit and anti-soul consensus view of materialist philosophers and scientists is based on *classical scientific materialism* that is disproved on a daily basis in high energy laboratories. Every parent should be asking why a flagrantly false philosophy is allowed to influence their children in the compulsory education system.

Young people attend educational institutions where they are encouraged to believe the false doctrine of materialism, which restricts their appreciation of themselves to the physical body as sole reality. Educated people who disbelieve in the soul may think they do so because they are better educated when in fact they may have been mis-educated.

The vortex theory offers a new direction in physics that supports belief in the soul. The understanding of how we can be both body and soul begins with appreciating that space and time owe their existence to the vortex of energy. This requires an understanding of Albert Einstein's theories of relativity, and the vortex theory can help us to understand how Einstein's theories work.

From Einstein's special theory of relativity we know there is no absolute, universal space and time. Einstein discovered that space-time and mass are all relative to the speed of light. This is easy to understand if mass is treated as the dense centre of the vortex of energy and space-time is considered to be its extension into infinity. Being different regions of the same system of spinning light, space, time and mass would both be relative to the same speed of light, spinning in the vortex of energy. This is detailed in *The New Vortex Theory* and *The Vortex Cosmology*.

The vortex theory explains away materialism, the idea that the universe is formed of things moving. No-thing exists that moves. Movement alone is the substance of everything. Movement spinning sets up the illusion of materiality. Everything associated with matter can be explained in terms of the vortex and everything is relative to the speed of motion in the vortex. In our world, that speed is the speed of light.

The force fields acting at a distance from particles of matter: electric charge, magnetism and gravity, are different expressions of the dynamic nature of the vortex of energy as it extends beyond our direct perception into infinity. They are all relative to the speed of light. They all belong to the same quantum continuum of spinning light. In *The Vortex Theory*[1] these forces are explained in detail.

The vortex of energy could be likened to an elephant with a massive body, peripheral ears, trunk and tail. The body would represent the massive centre of the vortex, the ears would represent the extending fields of electric charge and magnetism and the trunk and tail would be the peripheral extensions of space and time. A blind man feeling an elephant's trunk and then its ears may think they are entirely separate things. However, if he were suddenly to see the light he would realise they are not separate things but aspects of the same elephant.

Einstein's special theory of relativity helps us appreciate how each vortex of energy is a system of motion *relative* to the extension of space and flow of time, set up by every other vortex of energy. In the vortex theory this idea of vortex interdependence is presented as the *Universal Law of Love.*

Love your neighbour as yourself appears to be a fundamental law of the Universe. This law of love is the law of relativity without which there would be no space-time and every manifest thing would be unable to exist. *The Law of Love* suggests that the existence of every particle of vortex energy depends on the existence of all other vortex particles. This is because each vortex is the continuum of space and time for another vortex, as a form of motion, to exist and move in.

Quantum mechanics supports this idea. Werner Heisenberg had his epiphany when he watched a man walking under gas lamps in a street in 1925. The lamps threw a pool of light then were interspaced by inky darkness. As the man walked into a pool of light, he appeared, as he walked on into the dark, he disappeared. Heisenberg imagined the man disappearing out of existence and then coming back into existence again only when he interacted with the light. He then thought perhaps subatomic

1 Ash D. *The Vortex Theory* Kima Global Publishing 2015

particles, like electrons, only exist when they are interacting with photons of light or other particles of matter.

With Paul Dirac he developed the math of quantum mechanics based on the idea particles of light and matter exist only when they interact. Implicit in quantum mechanics is the idea that particles in the quantum world depend on each other for their existence.

The great success of quantum mechanics attests to the validity of the idea that particles of energy have a relational dependency on each other for their existence. No single particle of energy, as a system of waves or a system of spin, can exist as an island unto itself. In quantum mechanics and in the vortex theory this principle of total inter-dependence also applies to forces.

Newton's laws of motion show us that every action of a force depends upon an equal and opposite reaction. Every force in the Universe depends on something to act against. With no 'things' as such existing, each system of energy, as a bit of spin or a bit of wave, would rely on other bits of energy, as spin or wave, to act against.

Imagine a boxer. Every boxer needs another boxer to box against, or at least a punch bag. Boxing wouldn't exist as a sport if only one boxer existed in the world. People would dismiss him as the madman who kept punching air.

The continuum of vortex energy includes forces as well as space and time so each vortex of energy is an action relative to every other vortex as a reaction. This means the *Law of Love* is not just about mutual inter-dependence; it is also about mutual interaction. Reaction to action is a fundamental law of physics and both repulsion and attraction are aspects of vortex interactions. They are both expressions of the *universal Law of Love.*

Universal love is also about working with polarity. Polarity is part of the Law of Love. Everything and everyone has their place in the Universe and the interactions of opposites are fundamental to quantum reality.

We see this in art. Art is balance between space and form. Beauty in art is in the balance of these opposites and the energy in between them. It is also clear in dance. The dance is not the dancers it is a relational interaction between dancers

There has always been a polarity between those who believe in the soul and those who don't. Science cannot prove or disprove either point of view but in the new paradigm of super-energy a thesis unfolds, from Einstein's theories of relativity, which supports those who wish to believe we are simultaneously body, soul and spirit.

If space and time depend on vortices of energy existing on the planes of reality then space and time could not be the separation between the planes of energy and super-energy. If the difference between the physical and higher dimensional states is the speed of their energy then this could be what separates them. The idea that the planes of reality are separated by speed of energy rather than space and time is crucial to understanding the relationship between body and soul, and soul and spirit.

If space and time are dependant on the vortex of energy – or super-energy – while there may be space and time on each plane of reality there would not be space and time between the planes. This is because vortices setting up the space and time existing on each plane are 'on the plane' not 'between the planes'. This also applies to force fields. The law of gravity, for example, operates on each plane of existence but not between the planes. This thinking allows for the physical, hyperphysical and super-physical planes of reality to be in simultaneous existence –

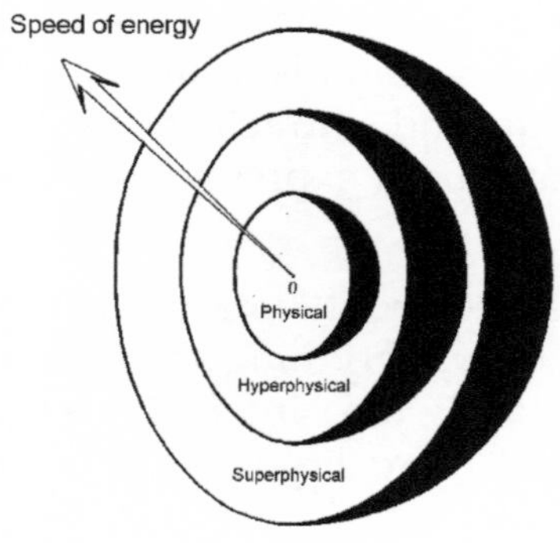

somewhat like sheets of paper impaled on a spike. This postulate is called: *The Principle of Simultaneous Existence of the Planes of Energy and Super-energy.*

The *principle of simultaneous existence* would enable a body of energy and a body of super-energy to coincide in the same 'here and now'. This principle would therefore allow for a body of super-physical energy – spirit – to interpenetrate a body of hyperphysical energy – soul – and these two levels of super-energy could coincide with and interpenetrate a physical body. In the new super-energy paradigm; the *paradigm of super-physics,* this idea is used to explain how an individual can be simultaneously a body, a soul and a spirit.

If forces are part of the quantum-vortex-continuum then while a soul might be subject to gravity and electromagnetic forces on the hyperphysical plane, it wouldn't be constrained by these forces on the physical plane so it could move through walls and levitate relative to physical reality.

Traditionally, people, like the Druids, believed we are simultaneously body, soul and spirit. They believed that the spirit is carried by the soul which interpenetrates the body while we are alive. The soul carrying the life-force of spirit then separates from the body when the body dies. Druids and other mystics appreciated that the soul then continues to live on in another realm of reality. If the soul is taken to be a body of hyperphysical energy and the spirit is assumed to be a field of superphysical energy, then the traditional idea of the simultaneous existence of body, soul and spirit would make sense in terms of the principle of simultaneous existence which allows for the coincidence of energy and super-energy.

In the past, belief in spirit and soul, as a psychic template that entered the body at conception and departed from it at death, was universal. Is it possible that some people saw the psyche or soul depart from a body with their own eyes? Is it possible that a minority of people could see things the majority did not? Maybe in the past the less aware majority accepted, with humility, what a more enlightened minority were able to perceive.

The anti-spiritual prejudice based on materialism has led to the marginalization of a minority group with inherited 'psychic' faculties that the majority of people lack. One group of people who experience discrimination, because of their abilities to perceive phenomena outside the narrow confines of the materialistic world view, are *spiritualist mediums* such as Doris Stokes, an ordinary woman with extraordinary powers of perception. Her autobiography, *Voices in my Ear*[2] began with her first memorable psychic experience.

When she was a child Doris had been awoken in the night by her parents rushing out to a fire in the neighbourhood. In her nightie and bare feet she followed her parents to the scene of the blaze. Pushing her way to the front, through the legs of the onlookers, she witnessed two firemen carrying the charred remains of an old man called Tom who died in the fire that night. What young Doris saw was Tom, alive and well, walking alongside the stretcher bearing his blackened corpse.

As an adult, after her father died, Doris had a vision of her baby, John Michael, dying in hospital. A few weeks later her baby was rushed into hospital with a blocked bowel. In her words this is what followed:

"I leaned against the side of the cot, my forehead wrinkled, wondering what to do. And then, I felt somebody watching me. I looked up quickly and gasped. My father was standing just inside the door.

"Doll," he said softly, "you know this isn't right. John Michael should be with us. He has to come back. At quarter to three next Friday I'll come for him and you must hand him over to me. Don't worry. I'll take good care of him," and then he vanished. One moment he was as solid as the furniture, the next I was looking at empty air."

Despite emergency surgery to save him, John Michael died the following Friday. This is Doris' story of how it happened:

"I could feel him going. I had to cuddle him one last time. Gently, gently I picked him up...I held him as tightly as I dared,

2 Stokes D. F. *Voices In My Ear*, Futura Publications, 1980

hardly taking my eyes from his face, memorizing everything. The soft baby smell, the pale silk hair, those long, long lashes. I must store it all in my mind. Time stopped. I gazed for hours, or maybe only seconds and when I raised my head, my father was standing on the other side of the cot.

"He didn't say a word. He looked steadily at me and then silently held out his hands. There was a long pause. I just didn't have a choice. Slowly, reluctantly, I passed my baby across, and at that very instant father took my son in his arms, I looked down and saw my little John Michael was dead."

Doris saw her deceased dad in a recognisable body and her own baby alive in his arms when the infant's lifeless body was still in the cot. Many people have had an experience of seeing and recognizing a relative who has died. Others, in an accident or during surgery, have seen themselves in a body they recognise as their own, looking down at their own physical body. Millions of people, during a *near death experience,* have seen relatives who they recognise. Either millions of people who claim to see souls are deluded or millions of people have actually seen themselves, or a relative, in a real soul body; a body in another dimension that is identical to the physical body. Maybe souls really do exist.

Maybe we have souls that survive death. Maybe they have the potential to reincarnate. Maybe the many reports of near death experience and out-of-body experience are based on fact. Maybe educated people who disbelieve in the soul are deluded by materialism. Maybe the time has come for a scientific revolution to bring back, into the mainstream, soul and spirit and real non-materialistic values.

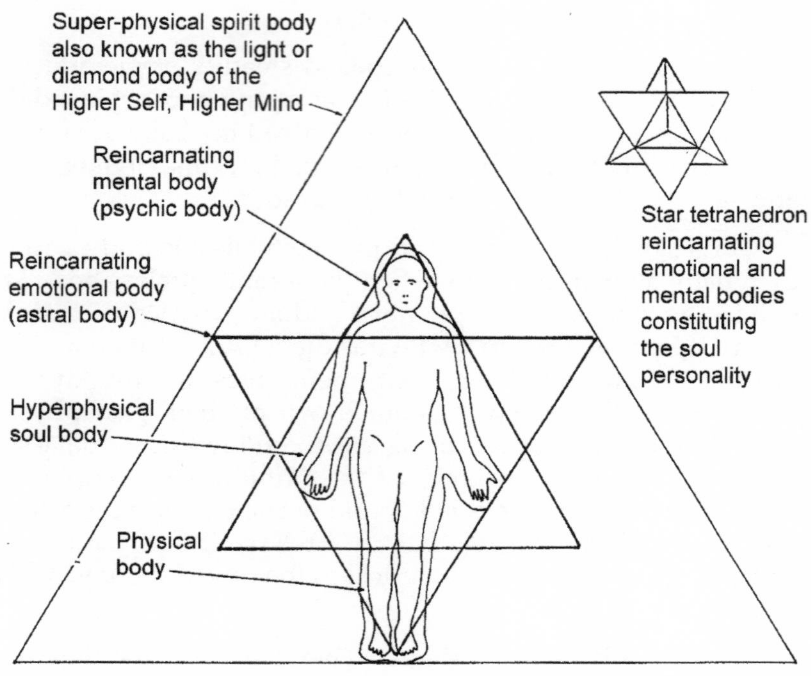

The many bodies of man

CHAPTER 6
THE TWO BODY DILEMMA

There are many people who have had a psychic, supernatural or spiritual experience. Many people believe in life after death and as for those who have had a near death experience they know death is not the end. Most of these people find it difficult to explain their experiences or account for their beliefs. Many don't like to speak about it for fear of ridicule. How is it possible to live after death? If we do, what form do we take? Do we have a body or are we just a cloud of consciousness?

Based on the assumption that super-energy worlds exist beyond the speed of light, what follows, in the new paradigm, is an attempt to help millions of people understand the nature of out-of-body or near death experiences, or a paranormal experience such as seeing a ghost. The proposal in the new paradigm of super-energy is that we have two bodies and not just one. One of our bodies consists of physical matter while the other is composed of hyperphysical matter. In my scientific story, the latter, called the *hyperphysical body,* is the soul body.

Each body – the physical and the hyperphysical – would exist in its own domain of space-time, one relative to the speed of light and the other relative to twice the speed of light. The *principle of energy speed subsets* – that lesser speeds are part of greater speeds – would determine that a physical body formed out of energy based on the speed of light would be part of the hyperphysical body, formed out of super-energy based on twice the speed of light, and not vice versa.

If we have a hyperphysical body housing the soul, we would be a soul with a body not a body with a soul. In the words of Thomas Aquinas, "The soul is not in the body, the body is in the soul."

The physical body may be the equivalent of clothing worn by a hyperphysical person. What you see in the mirror every morning may be the equivalent of a space suit that you, as a hyperphysical person, are wearing. It may be alive and you may be very bothered about it and how it looks but in the big picture it could count for little more than the clothes you put on in front of that mirror. If this is ringing true for you, the principle of energy speed subsets suggests that you are a spiritual being with a soul body and a physical body. You are here on Earth to have a physical experience. You are here to be grounded. You are not here to have a spiritual experience.

How is it possible to have a soul body and, if so, why is it invisible? The *principle of simultaneous existence* allows for a hyperphysical body to coincide with the physical body in the same 'here and now'. The *principle of energy speed subsets* suggests that the hyperphysical body – the soul body – would be invisible to the physical eye to all but those who have the brain type propensity or the gift to 'see between the worlds' and perceive hyper-physical energy.

A *two body dilemma* is experienced by many people. One of them was the atheist and sceptic Professor Howard Storm, who had an extraordinary near death experience.[1] Howard Storm found he was still in his body after he had left his body. This was a very bewildering experience for him.

In many near death or out of body experiences, people find themselves in a soul body that is identical to the physical body. A proposal to solve this *two body dilemma* is that the physical body acts as scaffolding for the hyperphysical body. As the physical body grows, it could be that the hyperphysical soul body is growing in parallel with it.

If scaffolding is used to build a house, once the house is built the scaffolding is removed. If the physical body acts as scaffolding for the hyperphysical soul body, then once the hyperphysical soul body is complete, the physical body can be dispensed with. That puts a whole new perspective on physical death. Death could be the equivalent of removing scaffolding when a house is complete.

1 Storm H. *My Descent into Death*, Clairview, 2000

It is the house which is important not the scaffolding. We live in the house when it is built and it is built to last, not the scaffolding. Logically the welfare of the soul body, which lives on continuously, should be considered more important than the wellbeing of the mortal physical body which dies. This could explain why in many religious and spiritual traditions, such as the Druids, more emphasis is placed on the welfare of the soul than the welfare of the mortal body.

The idea here is that the physical and hyperphysical worlds co-exist in parallel dimensions. In these worlds we are all growing two bodies in parallel, one out of vortices of physical energy and the other out of vortices of super-energy. Both bodies are atomic. The proposition is that atom for atom, molecule for molecule and cell for cell they grow alongside each other and are identical. The only difference between them would be the intrinsic speed of energy in every vortex subatomic particle within them. If the physical body determines the form of the soul body, physical life may determine its fate. The supposition that physical life determines the fate of the soul is the fundamental tenet of practically every religion and spiritual tradition.

The two body idea attempts to explain things that happen to people rather than explain them away. That is what a true scientist does. He or she uses science to explain things, not explain things away just because they don't fit with the consensus, materialistic point of view. As Richard Feynman said, "If your theories don't fit the facts it is the theories that are wrong, not the facts."[2]

When the body of physical matter dies, undeniable near death experiences suggest the identical body of hyperphysical matter simply steps out of it, like stepping out of clothing before going to bed at night. In the new paradigm of super-energy, death is viewed somewhat like a snake shedding its skin. Like a butterfly emerging from its cocoon, we emerge from the dead

2 Calder N. *Key to the Universe: A Report on the New Physics*, BBC Publications, 1977

physical body, amazingly beautiful and fully alive in a perfectly formed soul body.

As described in the Doris Stokes autobiography,[3] when Tom's physical body got burnt to a cinder, his hyperphysical body, that stepped out of the incinerated remains, was left bewildered and confused. Not knowing what to do it set off with the firemen and walked alongside the blackened corpse.

Which was the real Tom? To the amazed eye of little Doris the real Tom was walking alongside his corpse. No one else saw what she saw. Her father clipped her ear when he realised she had followed him out into the night, in her bare feet and nightie, and her mother thought she was mad when she tried to explain that Tom wasn't dead. But she had seen him alive and as large as life.

It is clear that most people don't see hyperphysical bodies wandering around. The exception to this is seeing a ghost. Ghosts can be explained in the new paradigm. In the physical world we cannot walk through a wall because the forces that hold our bodies together and make the wall solid also prevent our passage through it. However, while a super-energy being would be aware of a physical wall, as a part of its world, it would not be constrained by the wall. There would be no direct interaction between its super-energy and physical energy that holds the atoms in the wall together. A super-energy hyperphysical body could pass through the space in the physical atoms of the wall as if there was nothing there.

A ghost could be a person in a hyperphysical body. Those who claim to see ghosts see them as people in living bodies, sometimes in period dress, sometimes in ethereal white. Out of physical space-time the ghost could still be in the time – relative to us – when its physical body died.

Ghosts may be Earth bound souls that don't realise they have died. Earth bound souls are believed to haunt places familiar to

3 Stokes D. F. *Voices In My Ear*, Futura Publications, 1980

them, especially places where they might have gone through a tragic death. People alive today, who are resolved there is no continuity of life after death, might find themselves in a similar predicament after they die; a ghost trapped in its materialistic convictions.

Generally, we are not aware of super-energy beings because their energy speeds would be too fast for us to perceive. However, ghosts can be seen because of a strange phenomenon called *ghost-mist* or *ectoplasm* that will be discussed later.

Mediums, like Doris Stokes, can see people in hyperphysical bodies. This is because they can see into another dimension of reality with their psychic abilities.

Maybe our consciousness resides in the hyperphysical body and looks out into the physical world through our physical eyes, much as an astronaut would look out of the window of his space suit. It could be that our physical brains lock our consciousness into the physical continuum, that enables us to survive on planet earth, but a few psychic individuals have brains that allow them a greater degree of perception than most of us so they can be conscious of the hyperphysical as well as the physical.

The mainstream taboo on the experiences people have, when the constraints of their brains on perception are lifted, is tragic. When the doors of perception are opened, we may have an opportunity to gain a better understanding of ourselves and the Universe. That understanding could help us to lead less selfish and more meaningful lives. Belief in the soul can help us break the mould of fear and selfishness caused by the brain locking us in the survival mode of an animal. Loosening up our mindset to embrace the possibility of a continuity of life after death could help us prepare, by living a better life, for a more happy and meaningful outcome in continuous living. We may be wise to prepare for accountability just in case there is life after death.

The principle of energy speed subsets would allow that people living in the hyperphysical world could be aware of what is going on in the physical world. If our world is part of theirs, according to this principle, they would be able to see and hear us even if we cannot see or hear them. Mediums, like Doris Stokes, appear to have the ability to hear and see people residing in the hyperphysical dimension. In her autobiography she wrote how, in her sessions, they crowded into her open mind and seemed as keen to communicate with their relatives on earth, as their physical family members were keen to communicate with them.[4]

Attending a spiritualist meeting can be terribly boring as person after person receive seemingly trivial details from their loved ones 'on the other side'. But these details are vital to overcome natural skepticism. People need to hear intimate details of their lives that only a deceased family member or friend could know; something the medium could not know and the more trivial the detail the more unlikely it is the medium would know. Mediums travel constantly helping people realise physical death is not the end of life. They spread the truth that the energy of life is never destroyed and we are all destined for eternal life. They work tirelessly to help us appreciate that we are eternal beings; as Plato taught and Democritus denied.

Why are the cumulative mythologies of practically all cultures throughout the world excluded from science for the sake of an outmoded materialistic paradigm that is no longer valid in science? Surely, what most of humanity has considered undeniable truth, since the dawn of time, is worthy of scientific appraisal? As Graham Hancock wrote in his foreword to *Sun of gOd* [5] by Gregory Sams:

"For tens of thousands of years our ancestors lived in intimate proximity to what they believed were spirit worlds inhabited by non-physical supernatural beings, with whom they communicated. They saw spirits in mountains and trees and thunderstorms. Ocean and sky were alive with spirit. Fire and earth were filled to the brim with it. Entities and intelligences existed out there, inherent and immanent in every possible combination of seen and unseen realms.

"For about the last five thousand years various 'religions' have claimed hegemony over the ways we approach these spirit

4 Stokes D. F. *Voices In My Ear,* Futura Publications, 1980

5 Sams G. *Sun of gOd,* Red Wheel Weiser, 2009

realms and beings, often preferring to narrow down the focus to just one 'god'.

"During the past three hundred years, and particularly in the last hundred years, a form of thinking has arisen – it calls itself science – which in some cases makes a virtue of refusing to investigate such matters at all and in others proclaims that there are no spirit worlds, and no non-physical supernatural beings, and actively derides those who continue to believe in such 'myths'."

In fact science has no evidence of the nonexistence of spirit worlds and non-physical supernatural beings and it is wrong for scientists to allow us to think that is does. When a scientist asserts, for example, that there is no such thing as the soul and no possibility of life after death, he or she is not making a statement of fact based on empirical observations and repeatable experiments but rather a statement of unexamined prejudice based on personal beliefs about the nature of reality.

The body in which the consciousness of the individual was previously enshrined may have died, but the consensus materialistic proposal that consciousness cannot exist separately from the body – and thus has died with the body – is a metaphysical assumption, it is not an empirically demonstrable fact. The alternative proposition, which the world's wisest men and women have unanimously agreed upon for all but the past few hundred years, is that on death our consciousness passes on in some way on to another plane of reality; into another dimension of existence.

Maybe the reason why our ancestors saw spirits everywhere and in everything is because there are spirits everywhere and in everything. Is it that most of us are oblivious to many things that are going on around us? Think about it. We are aware of less than 1% of the electromagnetic spectrum with our physical eyes and less than 1% of the acoustic spectrum with our physical ears. Is our consciousness restricted, during physical life, to the limitations of the physical body?

This enquiry deepens. Does our consciousness reside in a hyperphysical body, which we inhabit but do not perceive? If we have two identical bodies, one that is subject to entropy, physical degeneration and death, and the other that is not subject to physical entropy, but lives on after death, then what the ancients perceived would make sense. With the new super-energy paradigm we can also make sense of some rites and rituals that come from more ancient times.

Catholics believe that when a priest recites the words, "This is my body and this is my blood", over bread and wine, during a service called the Mass, that the bread and wine become the body and blood of Christ. This rite, called the *Eucharist* is based on a covenant Jesus Christ established with his followers. On the night before he died, during a last supper with his small band of closest followers, he contracted that when they and those who followed them recited these words over bread and wine, his spirit would infuse the bread and wine in the way it inhabited his physical body.

The Eucharist can be accounted for in the new paradigm on the basis that when the words of consecration are recited, the *superphysical plasmic field* of Christ comes into coincidence with the appointed bread and wine according to the *principle of simultaneous existence.* Then, in accordance with the *principle of energy speed subsets,* the spirit of Christ could flow down the energy speed gradient into the person who receives the sacred food. The *Eucharist event* could occur for millions of people, simultaneously, on a Sunday morning, throughout the world, because, out of physical space and time, the spirit of Christ could reach an unlimited number of pieces of bread and cups of wine much as a single broadcast programme can reach an infinite number of radio or television sets.

In the Eucharist the Spirit of Christ channels the Universal Spirit into our hearts and our minds. This can also happen when pilgrims go to a sacred place. There the Universal Spirit – also called the Great or Holy Spirit – can overwhelm them according to the *principle of simultaneous existence* and fill their hearts according to the *principle of energy speed subsets.* Many people have a spontaneous *transcendental* experience of this nature at other times and in other places.

Transcendental experiences are not restricted to the Eucharist or any other religious rite or religious shrine. Angels or an individual's *higher spirit self* can channel the Universal Spirit into anyone who is open to receive it.

People do not have to be religious to have a mystical or spiritual experience. The more our hearts are open to the infinite mysteries of the Universe, the more likely we are, at some time in our lives, to receive a spontaneous gift of Spirit.

The Eucharist is an employment of the fractal principle that soul and spirit can coincide with a body of bread and wine as readily as a hyperphysical body and a superphysical field can coincide and animate a physical body of flesh and blood. By the same token the Earth, like the Eucharist, could have a coincident soul and spirit.

The idea the Earth has a hyperphysical body could account for the traditional idea of an *anima mundi* or *Earth soul*. Perhaps the human soul is a part of the Earth soul just as the human body is part of the Earth body and maybe our soul returns to the Earth soul after our physical body dies.

In *Timeaus,*[6] Plato described the anima mundi as an intrinsic connection between all living things on the planet, which relates to our world in much the same way as the soul (psyche) is connected to the human body. He wrote:

"The world is indeed a living being endowed with a soul (psyche) and intelligence... a single visible living entity containing all other living entities, which by their nature are all related."

The idea of an Earth soul was popular with hermetic philosophers like Paracelsus and other thinkers including Baruch Spinoza, Gottfried Leibniz and Friedrich Schelling, while the American writer, Ralph Waldo Emerson, was influenced in his writing by the Hindu concept of a universal soul; the *para-atman*. In more recent times the concept of an Earth soul would fit with the Gaia hypothesis, proposed by James Lovelock.

6 Plato and Jowett B. (translator), *The Complete Works*, Book House Publishing, 2017

Plato spoke of a reincarnating psyche acting as a body template. In the new paradigm of super-energy this would be treated as a reincarnating, superphysical *plasmic field* that acts as a body template, to direct the formation of both the physical and the hyperphysical bodies.

In our creation and formation, we could be fractals of the three universal planes, as proposed in the new paradigm; the superphysical, hyperphysical and physical. Formation in the same superphysical mould could help to explain *parallel differentiation,* how the hyperphysical and physical bodies could be identical, atom for atom and molecule for molecule. The only difference would be that the energy in the atoms and molecules would be based on different intrinsic speeds, represented by different Einstein constants of relativity.

If the hypothesis is correct, that the hyperphysical and physical bodies grow in the same super-physical crucible or womb, this might apply to all forms of life. It follows that every biological organism on Earth may have a hyperphysical counterpart that differentiates in a super-physical template. Every living thing would have soul and spirit carrying the life force within it.

The superphysical (spiritual) template over-shadowing the physical, and its hyperphysical (soul) counterpart, associated with every living organism – invisible to the naked eye but visible to the psychic eye – throws a new light on a beautiful and mysterious sentence from the 'Talmud' (an ancient compilation of Jewish moral and ethical debate): *Every blade of grass has its angel that bends over it and whispers. 'grow, grow'.*

The inherent urge and blueprint for the growth of organisms on Earth could be ascribed to an invisible template for life set up in the superphysical planes, which may well be the domain of our *higher self,* sometimes called the *god-self.* These superphysical (plasmic) electric-field blueprints for life, could account for gods, angels and devas.

Fairies and other elementals in nature that have featured, again and again, in all the world's mystical and folk traditions, could belong to the hyperphysical plane of reality. These nature spirits could be acting effectively through plant and animal souls that exist, perhaps like our own, in a parallel hyperphysical realm of reality.

The idea that a part of us lives on after death is held by people who feel they have 'been here before'. Many people have an intuitive belief in reincarnation. Circumstantial evidence of reincarnation in the scientific annals lends support to the idea that a part of us survives death and returns to live again.

CHAPTER 7
REINCARNATION

Belief in reincarnation has increased in the West, in part due to the interest in Yoga and all things Indian, but also because of the research carried out by a number of scientists, including Dr Ian Stevenson of the University of Virginia. Stevenson carried out extensive research into reincarnation over a period of twenty years. He collected over 2,000 cases studies in support of reincarnation and from 1960 onwards he published his findings in numerous journals and more than twenty books.[1] He emphasised that his findings did not prove reincarnation but provided a body of evidence to support those who wish to believe reincarnation is possible.

Dr Stevenson concentrated his researche on the testimonies of small children[2] and he took an interest in the relationship between birthmarks and past lives.[3] He chose mostly to study cases where children had spontaneous recollection of a past life that revealed details, which could be cross checked.

One case study was of a boy in France. He was born with a number of small birthmarks and as soon as he could speak he claimed the marks were left by bullets that killed him. As his speech developed he named the man who accused him of cheating at cards and then shot him. He detailed members of his family, the name of his girl friend and the village where he lived in Sri Lanka. His French parents found he was a difficult child

1 Stevenson I. *Twenty Cases Suggestive of Reincarnation*, University of Virginia Press, 1988

2 Stevenson I. *Children Who Remember Previous Lives: A Question of Reincarnation*, McFarland & Company, 2000

3 Stevenson I. *Reincarnation and Biology: A Contribution to the Etiology of Birthmarks and Birth Defects Volume 1*: Birthmarks 1997

because he insisted on eating with his fingers and demanding rice and curry. He wrapped a cloth round himself, fastening it like a sarong, broke into Sinhalese and attempted to climb trees in search of coconuts. Subsequent investigation revealed that a coconut picker had been shot for cheating at cards in the Sri Lankan village named by the boy, a few years before his birth in France. After the age of five, the memories began to fade and the boy grew up as a normal French child.

Dr Karl Muller also investigated cases where children spontaneously recalled lives.[4] He said it is not uncommon for children between the ages of two and four to speak as though they have had a previous existence. The problem with these cases is most children of that age find it hard to express themselves and are rarely taken seriously by their parents and by the time they are articulate the memories tend to fade.

A professor of psychology, Erlendur Haraldsson, started investigating reincarnation with Professor Ian Stevenson. He then went on to do his own ground breaking research with children who not only recalled episodes from past lives but remembered coming into incarnation. Over twenty years he collected over a hundred case studies.[5]

Another researcher, Dr. Frederick Lenz, investigated cases of people who claimed to remember a period of existence in a non-physical world. These people recalled death in a past life, passage through other worlds after death and subsequent rebirth. Lenz found the descriptions fitted closely to accounts in the Tibetan Book of the Dead. In his book *Lifetimes,*[6] Frederick Lenz recorded:

"I felt that all my life I had been dressed in a costume but I didn't know it. One day the costume fell away and I saw what I

4 Muller K., *Reincarnation: Based on Facts,* Spiritual Truth Press, 1970

5 Haraldsson E, Matlock J. *I Saw A Light And Came Here: Children's Experiences of Reincarnation,* White Crow 2017

6 Lenz F. *Lifetimes* Bobbs-Merrill 1979

really had been all along. I was not what I thought I was. All my life I had thought of myself as a body and a person...When I woke up in this world I realised I was not those things...It was like waking up from amnesia. I was overjoyed to be 'me' again. I had been all along but I had lost sight of it and thought I was the physical body. My body was only a thing I used for my life on earth. When it wore out I got rid of it."

According to Lenz many people have had experiences like this and identify the physical body as a prison that traps the real 'me' for the duration of life on earth. Many people, who have out-of-body, near-death or other-world experiences, identify more with being consciousness and mind than being a physical body. As their consciousness breaks free from the body, many of these people experience it as though waking from a dream:

"I found myself in a vast place. I felt as though I had come home. I had no apprehensions, fears or worries. I no longer remembered my former life on earth. Nothing existed for me but quiet fulfillment. I was not conscious of time in the usual sense; everything seemed timeless. I felt as if I had always been there. It was similar to the feeling I have when I wake from a dream that seemed very real only to realise it wasn't real, it was only a dream. That is how I felt. My former life on earth had been a passing dream which I had now wakened from."

According to Lenz, people who recall life beyond the physical body experienced it as coming into a greater level of consciousness. They speak of a passage between levels which fits with the prediction of higher speed states of energy setting up a series of realities in the Universe. From *Lifetimes*:[7]

"I did not have the sense of moving through space. Everything was consciousness and pure awareness...I moved through thousands of levels. On each level different souls were resting before being born again. The lower levels were much darker. I somehow knew that the souls on these levels were not as mature as those on the higher levels. Finally I reached a level that I was comfortable on. I stayed there. I sensed there were

7 Lenz F. *Lifetimes* Bobbs-Merrill 1979

many levels above the one I stopped at and that souls more advanced than I would go there."

Reports of consciousness surviving death are too numerous and universal to ignore. It makes sense that if consciousness can come into a physical body at conception and leave it at death, it should be able to repeat the process many times to benefit from different experiences, offered by varying cultures, in the procession of civilisations, down through the ages. Just as a child cannot learn all its lessons in a single day at school so it is that as an individualisation of consciousness we may need to incarnate in physical bodies a number of times to gain full benefit of the lessons to be learnt from physical life on earth.

The recall of many levels in the afterlife suggests that during physical life we are all mixed together, good, bad and indifferent but in the life after physical death it seems we are filtered according to our development, to a level most conducive to us where we can learn from our experiences and reflect on the consequences of our actions. Then it seems we come back again for more lessons in the terrestrial school where, rubbing up against others, we tumble, like gemstones against grit, to polish up a bit more.

If the physical body is considered to be analogous to outer clothing and the hyperphysical body to under garments, then the soul could be understood to be a super-physical psyche dressed in these layered bodies. The psyche as an individual personality incorparting the emotional *astral* and the mental *psychic* fields of plasma, would have been seeded into the cycles of reincarnation by the *higher spirit self* or *Angel-self.* The psyche is the *soul self* seeded by the angel-self that dresses in the two fresh body garments for each lifetime. It is this soul self, the psyche body template as described by Plato that reincarnates again and again, to evolve and grow over many lifetimes.

Many people have a gut feeling that death is not the end of life and that there is more to life than physical existence. The concept of soul and the idea that we somehow survive death is born out by millions of accounts following out-of-body experiences or near death experiences concurrent with advances in medical resuscitation.

CHAPTER 8
NEAR DEATH EXPERIENCES

Empirical evidence of life after death, and the existence of the soul, is coming from rapid advances in resuscitation medicine. These are leading to an ever increasing number of *near death experiences,* a term that was used by Dr Raymond Moody, one of the first to record extraordinary reports coming from patients who have been resuscitated.

In his book *Life after Life,*[1] Dr. Raymond Moody presented numerous case study testimonials in support of the idea of life after death, collected from people who had a near death experience. He described a typical near death experience:

"A man is dying and as he reaches the point of greatest physical distress, he hears himself pronounced dead by his doctor. He begins to hear an uncomfortable noise, a loud ringing or buzzing, and at the same time feels himself moving very rapidly through a long dark tunnel. After this, he suddenly finds himself outside his own physical body, but still in the immediate physical environment, and he sees his own body from a distance, as though he is a spectator."

Many people who have been through a death experience, in an operating theatre or a motor accident, report being out of their bodies, fully conscious, looking down on the scene of the operation or the accident:

"I saw them resuscitating me. It was really strange. I wasn't very high; it was almost like I was on a pedestal, but not above them to any great extent, just maybe looking over them. I tried talking to them but nobody could hear me, nobody would listen to me…"

1 Moody R. *Life after Life,* Bantam Books, 1967

Moody reported that many patients were aware of what people were saying and thinking about them:

"I could see people all around, and I could understand what they were saying. I didn't hear them audibly, like I'm hearing you. It was more like knowing what they were thinking, but only in my mind not in their actual vocabulary. I would catch it the second before they opened their mouth to speak."

In *Doors of Perception,*[2] Aldous Huxley wrote:

"...the brain does not produce mind, it reduces mind...each of us is potentially 'Mind at Large'. But in so far as we are animals, our business is at all costs to survive. To make biological survival possible, Mind at Large has to be funneled through the reducing valve of the brain and nervous system. What comes out at the other end is a measly trickle of the kind of consciousness which will help us stay alive on the surface of this particular planet... The various 'other worlds' with which human beings erratically make contact are so many elements in the totality of the awareness belonging to Mind at Large."

Near death experiences support Huxley's idea that the mind exists as a universal reality that is reduced rather than produced by the brain. Near death experiences suggest that when the brain dies, the cognitive mind returns from containment in the human body to the Universe at large. Only by accepting mind as a universal principle can near death and out of body experiences be understood within a scientific framework.

Many scientists and doctors, like the eminent neurologist, Sir John Eccles, have turned their backs on the materialistic approach that considers mind as limited to the brain:

"I maintain that the human mystery is incredibly demeaned by scientific reductionism, with its claim in promissory materialism to account eventually for all the spiritual world in terms of patterns of neuronal activity."

2 Huxley A. *The Doors of Perception,* Penguin Books, 1959

In the years since Dr Moody, accumulating testimonials coming from patients, who have experienced their minds when they were out of their body, support the position that the mind is reduced rather than produced by the brain:

"There was a lot of action going on, and people running around the ambulance. And whenever I would look at a person to wonder what they are thinking, it was like a zoom-up, exactly like through a zoom lens, and I was there. But it seemed that part of me – I'll call it my mind – was still where I had been, several yards away from my body…"[3]

Released from confinement in the brain the mind appears to be free to go wherever it wants in the world:

"When I wanted to see someone at a distance, it seemed like part of me, kind of like a tracer, would go out to that person. And it seemed to me at the time that if something happened any place in the world that I could be just there…"

Sam Parnia is a doctor whose specialty is resuscitation. In his book *Erasing Death,*[4] he explains that he doesn't start treating his patient's until after they have been pronounced clinically dead. He contends, the term near-death is inaccurate because, from his professional point of view, people actually die and have an *actual death experience* or *after death experience* rather than a *near death experience.*

Parnia said that half of the leading hospitals in Europe and America have resuscitation teams trained to bring people back to life after they have passed through clinical death. He said world wide, the accumulated reports of an afterlife, from people being brought back from the dead, now number in millions.

Parnia, and other critical care doctors, routinely interview patients asking if they had an experience between dying and being brought back to life. Parnia said that about ten percent of people recall a vivid, life changing experience of another world beyond the one we live in. At the time of publishing he had

3 Moody R. *Life after Life,* Bantam Books, 1967
4 Parnia S. *Erasing Death,* Harper One, 2013

collected about five hundred case studies. The recalls often involved having a life review, traveling through a tunnel, experiencing a loving being of light and seeing relatives before being drawn back into the body.

Most people who experience the other world report that it is similar to our own, but with more light and beauty. Only a few had an unpleasant experience of darkness and confusion. Most spoke of people in the other world having bodies like our own. They saw or were greeted by relatives, very often grandparents, who they recognised.

Physical bodies in our world grow with unique characteristics determined by the gene mix of random parentage. The formation of a physical body is unpredictable whereas the bodies of family members, who greet the people after they die, are predictable replicas – albeit younger – of their deceased physical bodies. That supports the idea that the physical body acts as scaffolding for the soul body – the hyperphysical body that goes to the 'other world' after death.

In his book Dr Parnia claims that science is divided over the extraordinary experiences recalled by people who have actually died and returned to life by the advances in resuscitation medicine. Attempts have been made to explain away near death experiences as hallucinations of a dying brain. Parnia points out that these do not stand in cases where the brain is already clinically dead before resuscitation begins.

In her PhD thesis on *The Near Death Experiences of Hospitalized Intensive Care Patients,*[5] based on a five year clinical study, Penny Sartori explained away all the usual explain-aways. In her book, *The Wisdom of Near-Death Experience,*[6] she makes it clear that near death experiences are not easy to explain away, if the subject is approached scientifically.

5 Sartori P. *The Near Death Experiences of Hospitalized Intensive Care Patients: A Five Year Clinical Study,* Edwin Meller, 2008

6 Sartori P. *The Wisdom of Near Death Experience,* Watkins, 2014

In *My Descent into Death,*[7] Howard Storm gives a graphic description of what happened when he died. In 1985, he had been left in a hospital bed in Paris following a perforated duodenal ulcer. An operation should have been performed within five hours but, as it was the weekend and the hospital had only skeletal staff, he was left for ten hours before surgery. The theatre staff were surprised when he came round after the operation as they hadn't expected him to survive.

Howard Storm had been in terrible pain for hours with no one but his wife Beverly, sitting by his side, holding his hand. Suddenly he found himself pain free standing up by his bed. He said the floor was cool beneath his feet, the light was bright and everything in the room was crystal clear. He was wide awake and not dreaming. Hospital smells assailed his nose. All his senses were heightened and alert. He clenched his fists to prove he was awake and not dreaming.

Howard Storm was aware of blood coursing though his body and the beat of his heart, pounding in his ears. He was very anxious and his mind was racing but he felt more alive than he had ever felt before. He spoke to Beverly but she didn't react. He started to shout at her but she still didn't respond, then he noticed a lifeless body on his bed under the sheet and was surprised at its resemblance to his own. It couldn't have been him because he was standing there looking down at it.

Storm's initial reaction was annoyance that someone had played a nasty trick and placed a wax model of him in the bed where he had been lying so ill and in so much pain, only moments before. He was very confused and screamed at his wife but she continued to ignore him. She just sat in her chair, weeping in despair.

Like Tom, in the Doris Stokes account,[8] Howard Storm found he was alive standing beside his lifeless physical body. He was confused and anxious because, as a university professor and a

7 Storm H. *My Descent into Death,* Clairview, 2000
8 Stokes D. F. *Voices In My Ear,* Futura Publications, 1980

confirmed atheist and sceptic he was not prepared for the continuity of life after physical death in a body identical to the corpse he had just stepped out of. His experience could be imagined somewhat like an astronaut stepping out of a cumbersome space suit. He was still in the hospital, and he was still in a body with all its senses but they were more heightened as though he was released from being damped down. He could see, feel, hear and smell everything and see and hear other people in the hospital, but they could not see or hear him.

Storm's account fits the two body hypothesis. His experience of a two body dilemma was common to many accounts given by people who have had a near death or out-of-body experience.

Testimonials of near death or out of body experiences are not accepted by materialist scientists and philosophers, not because they are not true, but because they do not fit with what these intellectuals believe to be true, according to their materialistic *quasi-religious* mindset. Truth is what the people, who control politics, education and the media propound. Truth has more to do with what scientists believe than with what they discover. In *The Structure of Scientific Revolutions,*[9] Thomas Kuhn pointed this out when he wrote:

"Truth has as much to do with the consensus of scientists as to the outcome of experiments."

People in power write history in their favour. They control beliefs. In the past it was the religious establishment that had the power over people's minds so priests were in control of the truth. Nowadays it is predominantly the university establishment of scientists and philosophers that determines what is true and what is false. None of this has anything to do with what is actually true. So what are we to believe? The answer is very simple. We don't have to believe what other people tell us, or what they believe and that includes what I write. We are all free to believe according to our own experience and enquiry. There is a place inside us where we feel intuitively what is true or false.

9 Kuhn T. *The Structure of Scientific Revolutions,* University of Chicago Press, 1962

We all have an intuitive knowing, a gut feeling for the truth; and that is usually what is right for us to follow.

What is the purpose of life? Is there even a purpose? Many people don't believe that there is a meaning or a purpose to human life. However, a scientist in the 18th century went through a life changing near death experience that led him to propose that there is a profound purpose for human life. On his visit to a higher dimension he was told that the planet Earth is effectively a *womb of angels*.

CHAPTER 9
THE WOMB OF ANGELS

Emanuel Swedenborg was a well educated, widely traveled Swede who went through a profound spiritual awakening in 1745. It occurred during a health crisis from which he nearly died. He said he had been to another dimension where a heavenly guide gave him revelations for humanity. After his crisis, he continued to have spiritual visions and dreams. Swedenborg had already established himself internationally as a scientist, inventor, mathematician and philosopher so he was taken seriously by a great number of people although just as many thought he had gone mad.

The teachings of Swedenborg,[1] had a profound influence on Blake, Goethe, Emerson, Dostoevsky, the French Symbolists, Kant and Jung. The essence of his message was that humans exist simultaneously in physical and spiritual worlds. After death memories of the physical world fade whereas those of the spiritual world survive. He said Heaven is much like the earth except that people live in spiritual bodies rather than physical bodies and they are able to enjoy pleasures, including sex, but life there is comparatively mundane.

Swedenborg also taught that Hell exists but there are no devils or Satan there. The core of Swedenborg's revelation is that God, Heaven and Hell exist within us. He was told that we should ditch doctrines like the Trinity and the idea that Christ died on the cross to atone for the sins of humankind. Moreover, the teacher he met on the spiritual plane told him that salvation occurs through personal striving to live a spiritual life.

1 Swedenborg E., *A Swedenborg Sampler: Selections from Heaven and Hell, Divine Love and Wisdom, Divine Providence, True Christianity, and Secrets of Heaven,* Swedenborg Foundation Publishers, 2011

Emmanuel Swedenborg devoted the last twenty seven years of his life to teaching and writing. His most famous books were *Arcana Ceolestias*[2] and *Heaven and Hell.*[3] One of the most significant points in Swedenborg's heterodoxy is that:

"The human race is the basis on which heaven is founded and angels arise from the souls of humans."

From *The World of Angels,*[4] :

"Contrary to orthodox theology, Swedenborg believed that angels are not created in heaven by God but arise from the souls of deceased human beings. In the Swedenborg celestial hierarchy, a soul's place in angelic society is determined by his or her beliefs and sensibilities as an earthly mortal."

This teaching of Emanuel Swedenborg has profound implications. If Swedenborg is right, then in effect the Earth would be a *womb of angels*. Our planet would be a place where angels are born and gestated. We may be baby angels.

The powerful idea that the Earth is a womb of angels casts a positive light on death as a time of birth. Just as when we are born into the physical world we passed through a tunnel – the birth canal – into the light of a new world, so when we pass into the next world, from this world, we pass through a tunnel into the light of the new world. This is corroborated by testimonials from thousands of near death experiences.

However, I believe it is through ascension (which I will discuss later) rather than death that a new angel is born. Swedenborg endorsed the view that we become an angel and join the celestial choirs of heaven when we ascend, not when we die. My belief is that when humans die they are born into

2 Swedenborg E. *Arcana Coelestia,* Forgotten Books, 2008

3 Swedenborg E. *Heaven and Hell,* Swedenborg Foundation Publishers, 2010

4 Penwyche G. *The World of Angels* Bounty Books 2009

another level of the Earth from whence their soul personalities reincarnate back onto the physical plane again.

I consider that the hyperphysical world (the fourth dimension) we enter after death is but a phase between many physical incarnations required to prepare for our birthing as an angel. I hold this belief because I do not believe a single lifetime is sufficient to become a fully fledged angel. In my new paradigm I consider *the womb of angels* to incorporate the hyperphysical plane as well as the physical. I believe humans are born as angels when they ascend into the super-physical from both the physical and hyperphysical planes of the Earth.

If we understand the hyperphysical plane as being integral to the Earth – the *Earth Soul* or *anima mundi* – then that is where we would most likely journey to after physical death. However, following the death of our physical carrier bodies we would not necessarily enter the superphysical planes of the Universe (the fifth dimension and beyond) where angelic and divine intelligences reside.

If, after an exceptional lifetime, someone is ready to move on and ascend to the fifth dimension and higher worlds, it may be in the wake of many incarnations on Earth gestating and training as an angel. It makes sense that we would need more than one lifetime to prepare us for continuous living in the Universe at large.

Each lifetime on a physical Earth could be likened to a day at school where we receive the lessons we need to widen our experience and understanding and evolve our sentience to that of an angel. Death of a physical incarnation would be the equivalenty of the 'end of a day' at angel school when our hyperphysical soul body returns to the hyperphysical realm. That is where we would go for rest and integration.

We could then drop the cloak of identity we adopted on Earth and integrate the experiences of physical incarnation. We could also prepare for our next lifetime; our next day at angel school. We could review the upcoming syllabus and ready ourselves for the experience of a new hyperphysical and physical body.

During this hyperphysical interim, which is part of the *Earth Womb of Angels*, we would be able to access a vantage point outside of the conflict, duality and separation experienced during physical incarnations on Earth. A clear-sighted perspective of our relationship to the unity and integrity of the Universe would be vivid and our appetite for advancement might be stirred, during our review of a lifetime spent on Earth, and impel us towards our next life in physical incarnation.

After physical death, the fruits of the physical life are pressed through the hyperphysical wine-press. Each soul finds itself attracted by resonance, according to its nature and desires, to a heavenly or hellish situation in the hyperphysical world. These polarity situations are self imposed. As Emanuel Swedenborg said, "Heaven and hell are from the human race."

Heaven and hell present opportunities for a soul to review actions chosen during its lifetime and crucially, to taste and experience the consequences of these actions – some sweet, some bitter and some rotten – reverberating through the web of life. The experience of joy or sorrow felt by a soul in the 'after-life' is determined by how much that soul affirmed or negated itself.

Joy and sorrow, heaven and hell, are hi-fidelity states of the soul. They provide necessary extreme experiences. One isn't necessarily better or worse than the other. Both polarities provide rich feedback on the condition of one's soul (the Atman) in relationship to, or in divergence from, its highest spirit state (the Para-Atman). This is also referred to as the 'Higher Self', the 'Divine portion', the 'Angel-self' or the 'I AM' presence.

Finding oneself in heaven or hell is where one can best and most honestly review one's learning curve. Both situations offer the reincarnating spirit an opportunity to set about cultivating a capacity for greater soul cognition. This is in line with the utmost intelligence of the Universe; the force that propels all life towards greater complexity, resolution and refinement. As Julian of Norwich said, "Hell is a greater depth of God's love."

On both the physical and hyperphysical levels of the Earth, for souls that have raised their thoughts and actions to the standard of Universal love and fellowship, a heavenly experience is open to them. Conversely, the opportunity for a hellish experience is available to souls that refuse to love. Contrition or profound sorrow and grief, as a consequence of thoughts and actions that caused pain, anguish and despair to others or oneself, can help elevate the soul from a state of hell to a state of compassion in both the physical and hyperphysical levels of reality.

On both the physical and hyperphysical planes of the Earth, heaven is always available to souls experiencing hell if they can raise themselves, in their minds, from darkness to light, from despair to faith, from recrimination to forgiveness, from hate to love, from pride to humility and self-righteousness to contrition. Despite the fact that it is extremely rare for a soul to achieve this on the hyperphysical plane, if they have not done so on the physical, the opportunity is always there for them to do so.

No entity keeps a soul in hell apart from the soul itself. Souls choose to remain where they are. In this light, heaven and hell are not places of reward or punishment where souls go 'after God's judgment' but situations related to a soul's condition. After death the soul would be propelled by its own deepest instinct for learning and evolution to join the *frequency zone,* in congregation with other souls of like mind. The downside to this arrangement of partitioning of souls into 'like minds', makes it less likely for a soul to undergo a radical change of mind. That is what incarnation on the physical plane is intended to achieve.

Humans on different 'wave-lengths' are mixed on the physical plane to provide reincarnating souls the opportunity to learn from one another. They are then partitioned on the hyperphysical plane to enable them to integrate their learning with others who are most akin to support them in this.

Another way of viewing this is that after death a universal law of attraction pulls the soul towards a polarity of experience based on resonance according to the energy signature of the soul. The soul signature is based on accrued imprints of thought and feeling following completion of a physical incarnation. The soul signature determined during physical life sets the destination for the soul in the hyperphysical levels of reality, which serves the soul's best interest for internalized learning.

The fires of hell are not an external phenomenon but are rather an opportunity for learning through the purging fire of internalised frustration anger, grief, anguish and despair. In the hyperphysical realms of energy the soul doesn't have the dulling physical body. It is in a heightened state of awareness of thoughts and emotions engrained during physical life that causes either ecstasy or anguish.

In the hyperphysical realms, souls are aware of each other's thoughts so they cannot hide. They have to face themselves and others and live with themselves and with others as they truly are. This process, whether heavenly or hellish, is more advantaged by being with others of a similar disposition.

Heavenly or hellish experiences are experienced as the space of light or dark existing within the soul. The experiences of heaven or hell are merely intensified on the hyperphysical plane by removal of the physical 'damping' body and physical distractions.

If we are spirits designated as infant angels, if we are gods in the making, it makes sense we would need to go through tough training interspaced with uncompromising periods of life review to prepare us for eternal life in the Universe at large. If this world view is correct there would be a purpose to suffering. Neem Karoli Baba, a famous Indian sage and Guru for many in the West, including Steve Jobs, said, "I love suffering; it brings me so close to God".

It is for our ultimate good that we all have to go through pain and sorrow in the *Earth Womb of Angels* throughout our many incarnations in physical body, and residual interludes in soul body between them. The Lebanese poet Kahlil Gibran famously captured the redemptive value of difficult experiences in *The Prophet:*

Is not the cup that holds your wine the very cup that was burned in the potter's oven?

And is not the lute that soothes your spirit, the very wood that was hollowed with knives?

When you are joyous, look deep into your heart and you shall find it is only that which has given you sorrow that is giving you joy.

Pupils at school may hate the school system but their protests fall on deaf ears. We may hate our bodies and begrudge our lives but we have no choice but to endure physical life and learn what we can from our experiences because even suicide is no escape.

Suicide lands souls in an even deeper and darker self-consumed hell than the one they tried to escape from. This is clear from testimonies of near death experiences recorded by people who have been resuscitated after attempting suicide.

In the words of German poet and student of Rudolph Steiner, Christian Otto Morgenstern (1871-1914):

"Our desire no more to suffer causes only new pain,
Thus will you never shed your garment of sorrow,
You will have to wear it until the last thread,
Complaining only that it is not more enduring,
Quite naked must you finally become,
Because by the power of your spirit,
Must your earthly substance be destroyed,
Then naked go forward in only light enclosed,
To new places and times, to fresh burdens of pain,
Until through myriad changes a god so strong emerges,
That to the sphere's music you your own creation sings."

There appears to be no escape from suffering in the *Earth Womb of Angels*. Everyone has problems to contend with. No one is exempt. Only when we learn to treat others as we would wish to be treated ourselves, only when we support each other in times of difficulty, only then will our own suffering be reduced.

The lesson of *compassion* is the single most important thing for humanity to learn on the Earth plane. We find love by loving and caring for others.

To quote William Blake:

"Seek love in the pity of other's woe, in the gentle relief of another's care, in the darkness of night and the winter's snow, in the naked and outcast, seek love there."

The golden key to liberation from suffering associated with the experience of duality on Earth is to literally burn with compassion for all of life. It seems to me that we are here to learn that when we give ourselves in selfless service and act out of unconditional love then our own footprint of suffering as well as the suffering of others is mitigated and misery is replaced by joy, even in the most dire of situations.

To quote Albert Einstein:

This delusion is a kind of prison for us, restricting us to our personal desires and to affection for a few persons nearest to us. Our task must be to free ourselves from this prison by widening our circles of compassion to embrace all living creatures and the whole of nature in its beauty.

The children in war torn Syria expressed, on worldwide television, the joy they discovered when they looked after each other. As they surrendered to their suffering, they discovered joy in acceptance of their fate as the will of their God.

Diamonds, the most precious and admired stones on the planet, probably the strongest material on earth, are forged through exposure to intense heat and pressure. Similarly, history reveals that the human spirit grows strongest through adversity and difficulty and is weakened by comfort and leisure.

Just as infants receive vaccinations to help them withstand debilitating diseases in their future adult life, so the universal experience of life on Earth is one driven and directed by fear, with a view to ultimately eradicating fear. This existential process of wrestling with fear and eventually overcoming it is strikingly similar to the transformation of coal into diamonds.

Fear is the most crippling of companions for any wayfarer. However, the journey through lives of fear and the ultimate triumph over fear is the evolutionary leap that qualifies a soul for angelhood.

Gautama Buddha identified the pathology responsible for protracted human suffering. He said that suffering is caused by an untrained mind being eternally caught in a frenzied chase between craving pleasure and avoiding suffering.

The solution to this dis-ease he taught was to train the mind through meditation and to cultivate equanimity. The essential nature of our sovereign soul, he revealed, is to be free from bondage to pleasure or pain. Rudyard Kipling crystallised this perennial wisdom in his famous poem *If*:

"...If you can meet with triumph and disaster and treat those two imposters just the same."

Souls are especially challenged by incarnating in the highly materialistic civilisation of the twenty first century, where the overarching cultural message is to pursue pleasure and avoid pain and suffering, and that the source of ultimate value is gratification of desire. For so many people life is a dedicated campaign to screen off suffering with comforts and pleasures. For many greed is considered good and money is god.

The root of the materialistic mindset is fear in the form of insecurity and fear of uncertainty. Fear is why many people attempt to dominate and control others and control the course of their lives. Fear can be embraced, transmuted and overcome when we move into the unknown and choose to live with courage and trust that all our needs will be met. This is called, *living fearlessly*. It is also called 'the way of faith' or 'living in faith', I prefer to call it 'playing with the Angels' or 'playing the game of trust'.

Fear must be confronted, overcome and embraced by the soul, for fear guards, like a fire breathing dragon, the greatest spiritual treasure any wayfaring soul can find on this planet which is the ability to live fearlessly. Fearless souls are the most likely to step into the light of ascension and into angelhood.

BOOK VI

MIRACLES & MAGIC

There is such an order in this intricate vast mystery, a sort of preordained pattern in everything. It's like we all have our destiny and our potential for vibrant health so it doesn't matter what comes along to change it or try to destroy it, if we have faith in the infinite re-creative energy real healing can happen.

Anna Mary Ash

CHAPTER 1
PSYCHIC SURGERY

Psychic surgery is an extraordinary procedure through which many thousands of people have experienced dramatic healing. Materialists decry psychic surgery as a hoax. They accuse psychic surgeons of fakery and say the patients have been duped. This disparaging attitude stems from the inadequacy of the materialist paradigm. The materialistic frame of understanding they are embedded in fails to account for supernatural and paranormal phenomena.

A new frame of understanding, emerging from the vortex theory, is based on the assumption that super-energy exists beyond the speed of light. This can account for supernatural and paranormal phenomena in a scientific context. I call this frame, *super physics*. Within this new scientific frame of understanding there is a simple way of explaining psychic surgery and other seemingly 'miraculous' phenomena.

Many miracles can be explained in terms of *super-energy resonance*. Super-energy resonance occurs when an intense beam of super-energy is applied to matter. Resonance then occurs between the energy in the matter and the super-energy in the beam. This resonance causes the energy in the vortices and waves of the matter to undergo acceleration of its intrinsic speed value and come into equilibrium with the intrinsic speed value of super-energy in the beam. If the super-energy beam is switched off, the energy reverts to its original 'natural' speed value. There are therefore, two modes to the super-energy resonance process.

Energy ascension is the mode in *super-energy resonance* when the speed of energy in subatomic 'vortex' particles and 'quantum' wave particles is *accelerated* to speed values exceeding the speed of light. In this procedure matter and light would be ascended or *evaporated* from the physical state to the hyperphysical or superphysical states of matter and light. In this process the matter or light would vanish. They would not 'go anywhere' as such but as their energy is accelerated to super-energy they would no longer be in physical space and time so they would no longer reflect physical light or interact with physical matter.

Super-energy descension is the mode in *super-energy resonance* when the speed of super-energy in hyperphysical or super-physical 'vortex' subatomic particles and 'quantum' wave particles are decelerated to the speed of light. In this procedure hyper-physical or superphysical matter and light would be descended or *condensed* from the hyperphysical or superphysical states of matter to the physical states of matter. In this process the newly condensed matter or light would appear in our world. On deceleration of the super-energy to physical-energy they would be in physical space and time so they would reflect physical light and interact with physical matter.

The strange phenomenon of psychic surgery can be explained in terms of *super-energy resonance*. In 1959, psychic surgery came to public attention when Ron Ormond published *Into the Strange Unknown*.[1] He called the practice 'fourth dimensional surgery,' and wrote:

"We still don't know what to think; but we have motion pictures to show it wasn't the work of any normal magician, and could very well be just what the Filipinos said it was — a miracle of God performed by a fourth dimensional surgeon."

In psychic surgery, which originated in the Philippines, a psychic surgeon – who never received medical training – runs his hand over the skin of the patient, as he does so a hole appears as though he used a scalpel, only no scalpel is used, and there is no pain or bleeding. The psychic surgeon then plunges his hand into the hole – sometimes with a blunt knife and never with an anaesthetic – and performs an operation.

1 Ormond R., *Into the Strange Unknown* Esoteric Publ., 1959

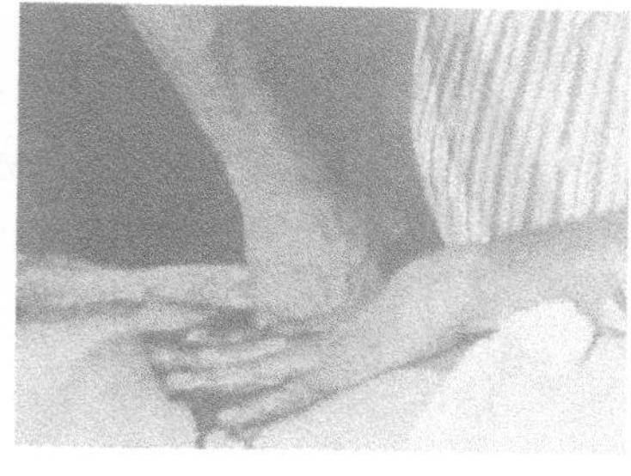

The psychic surgery may be the removal of a cancerous tumour, a blood clot or some other obstruction. That is when some bleeding may occur. At the end of the operation, the psychic surgeon – often illiterate – then passes his hand over the hole and it disappears, leaving the body and skin intact as it was before the operation, without stitches, pain or a scar.

John of God, working in Brazil, is the most famous psychic surgeon of our time.[2,3] He started out as a farmer who gave up three days a week to practice psychic surgery, free of charge. Now he works full time as thousands of people come to him from all over the world. One of them was the mother of my younger son.

In the frame of *super physics*, psychic surgery is explained in terms of a 'laser like' application of *super-energy resonance*. If an intense beam of super-energy is applied to a specific part of a patient's body it would cause the energy in every vortex and wave particle in every atom and molecule in that region of the patient to *resonate* with the super-energy in the beam. During this resonance the intrinsic speed of the energy would accelerate beyond the speed of light to match or come into equilibrium with

2 Cumming H., *John of God: The Brazilian Healer Who's Touched the Lives of Millions*, Atria Books, 2007

3 Meliana M. *John of God: A Guide to Your Healing Journey with Spirit Doctors Beyond the Veil*, Blue Leopard Press, 2014

the speed of the super-energy in the beam. Every particle of energy in every atom, molecule and cell in that region would then ascend from the physical to the hyperphysical state of matter. That would cause the tissues in the beam to disappear out of physical space. Switching off the super-energy resonance beam would reverse the process enabling the super-energy to revert to physical energy. The energy in the flesh would then descend from the state of hyperphysical matter to physical matter. The tissues would then reappear in physical space again.

The question is how the super-energy resonance would be applied. It could be, as the Filipinos said, that a fourth dimensional (hyperphysical) surgeon was operating the super-energy resonance technology. We could call this surgeon a *hyperphysical operator*. As suggested by the Filipinos, this could be a discarnate surgeon working, from the hyperphysical fourth dimension,[4] in collusion with the psychic surgeon operating in the physical third dimension.

The idea is that the hyperphysical operator uses a *super-energy resonance* beam to accelerate the speed of energy in the specific region of the patient's body, to be operated on. This is indicated by a sweeping hand motion of the psychic surgeon – much as a nurse indicates to a surgeon where to cut by drawing a line on the skin of a patient with a marker pen. In the case of psychic surgery, the flesh would be parted, not by a scalpel in the hands of a physical surgeon, but by an intense *super-energy resonance beam* – just like a laser beam – directed by the unseen hyperphysical operator.

To reiterate, the *super-energy resonance beam* could be likened to a laser beam, only it would be a beam of intense super-energy not physical energy. Vortices and waves of physical energy coming into the beam would resonate to the speed of super-energy within the beam. As the beam is applied to the patient's flesh, all the cells focused on by the beam would vanish beyond the speed of light. The physical-energy in that region of

4 Ormond R., *Into the Strange Unknown* Esoteric Publ., 1959

bodily matter would be changed into super-energy. As the speed of the energy in the flesh accelerates beyond the speed of light, the flesh would slip out of visibility into the fourth 'hyperphysical' dimension. In the *hyperphysical micro-zone,* set up by the *super-energy resonance beam,* the tissues would literally vanish out of physical space and leave an 'apparent' hole in the patient's body into which the psychic surgeon could plunge his hand to perform an operation. It is that simple.

In this extraordinary procedure, the physical tissue that becomes invisible does not go anywhere and is not disturbed by surgical incision. It merely passes beyond third-dimension human perception and physical interaction. In the course of the *super-energy resonance* procedure there would be no change in molecular structure of the tissues involved. They simply become a micro-zone of super-energy.

Physical blood flowing through the vessels within the operational zone would be temporarily suspended out of physical space. In their passage through the micro-zone of hyperphysical super-energy, the blood cells would correspondingly change status to hyperphysical blood.

On leaving the micro-zone of super-energy, the patient's blood would revert to physical blood without evidence of any wound, scar tissue or breakage of blood vessels. As it moves in and out of the hyperphysical space-time zone, the 'ascended blood' would continue to flow through the operational zone, even passing through the psychic surgeon's physical hand, due to its inter-dimensional status. This would be in accordance with the *principle of simultaneous existence* which allows for super-energy to coincide with physical energy but be totally unobstructed by it.

Any bleeding that occurs would be as a result of the operation performed by the psychic surgeon in the underlying tissues below the hyperphysical operational zone. Those tissues and organs would still be in physical space.

On completion of the operation, the super-energy resonance beam, which set up the hyperphysical micro zone, would be switched off by the hyperphysical operator. The vortex and wave particles inside the atoms and molecules in all of the cells of patient's flesh, having been ascended, would immediately revert to the physical state. The muscle and skin flesh would then reappear in physical space. The patient would appear to be instantly healed and a miracle would have been deemed by onlookers to have occurred.

This is a relatively simple account for psychic surgery but any attempt to explain such paranormal phenomena or understand miracles is impeded by the divide between science and religion. Religious people proclaim miracles as acts of God while scientists proclaim them as hoaxes. If superstition can be superseded by an appreciation of the potential of *super-energy resonance* then miracles and suchlike supernatural phenomena can be reasoned inside a new scientific framework.

The extraordinary abilities of one man in particular can be explained in terms of *super-energy resonance*. This *man of miracles* was renowned world-wide for his ability to heal miraculously and to manifest, apparently out of nowhere, expensive jewelry, watches, pendants and sacred ash called *vibhuti*, in a way that persistently confounded the scientific establishment.

CHAPTER 2
MIRACLE MAN

Born Sathya Narayana Raju (1926-2011), Sathya Sai Baba, as he became known, was an illustrious Indian guru, miracle worker and philanthropist who resided in Andhra Pradesh, India.

The display of miracles, throughout Sai Baba's life, led to an enormous following in the West as well as in India. He was famous for his ability to manifest valuable jewellery out of thin air for his devotees,[1] including rings, bracelets, pendants and expensive watches. Those lucky enough to receive one of his lavish gifts treasure them to this day.

Sathya Sai Baba

Many recipients of Sathya Sai Baba's gifts have had them authenticated by jewellers and the gold has always been real gold and the gem stones genuine. Sai Baba may have died but his rings and watches live on, worn on the fingers and wrists of numerous individuals.

Sceptics said that Sai Baba was a very clever magician. However, it is very unusual for magicians to give away their props, especially expensive items of jewellery.

When asked how he performed his miracles, Sai Baba explained that the rings, watches and ornaments existed in his *Sai Stores*. All he had to do was visualize them in his mind and they would appear in his hand.

1 Haraldsson, E. *Appearance and disappearance of objects on the presence of Sathya Sai Baba.* J.Am.Soc.Psy.Res., Jan. 1977

The ability to perform miracles – a psychic ability called *Siddhi* in India – was acquired by Sathya quite suddenly when he was fourteen. In March 1940, after recovery from a loss of consciousness, for twenty four hours following a deathly scorpion sting, he had a spontaneous spiritual awakening. In the weeks after the near-death-experience Sathya began a mission as a spiritual teacher, or *guru,* and in May 1940 he performed his first miracle.

When the fourteen year old awoke from the twenty-four hour coma, his family found his behaviour very strange. He would alternately laugh and cry uncontrollably. Then he quoted long passages of Sanskrit philosophy and poetry that he had never learnt and he described in vivid detail distant places he had never visited. For two months his family called on the help of an expert in devilry to exorcise the demon they believed to have taken over their son, but to no avail.

When, on the morning of May 23rd 1940, the boy performed his first miracle, in front of his family and some people from his village, the attempts at exorcism were immediately abandoned. Sathya had suddenly 'materialised' flowers and sugared candy as if out of nowhere.

Sathya was asked "who are you?" by someone amongst the amazed gathering. The boy replied calmly but firmly: "I am Sai Baba". He explained he had been 'reincarnated' through the prayers of an Indian sage, one of India's most revered modern saints, Sai Baba of Shirdi, who had died in 1918.

Shirdi Sai Baba

Shirdi Sai Baba, as he was also known, had preached the importance of self realisation and criticised love of perishable things. His teachings concentrated on a moral code of love, forgiveness, helping others, and practicing charity, contentment, inner peace, and devotion to God and guru. Shirdi Sai Baba condemned distinction based on religion or caste. He was neither Muslim nor Hindu, as religious distinctions were of no consequence to him. His teachings combined elements of Hinduism and Islam.[2]

The villagers and Sathya's family, struggling to accept the possibility that Shirdi Sai Baba had 'come again', gathered around him. "Show us a sign", they demanded. With a quick and unexpected gesture Sathya threw a bunch of jasmine flowers he manifested from thin air onto the floor, they spelt out, in the language of the village: Sai Baba.

Sathya's newfound spiritual life accelerated at a rapid pace. By 1944, *Sathya Sai Baba,* as he now called himself, had attracted a considerable following and his devotees had built a Mandir (Hindu temple) near his home village of Puttaparthi in Andhra Pradesh. In 1948, they built an ashram for him, which was completed in 1950. By 1954, Sai Baba had established a small free general hospital in the village Puttaparthi, which was the beginning of his life-long philanthropy.

The fame Sathya Sai Baba garnered from his miracles and his ability to heal had spread far and wide across India and by the 1960's he had become well known in the West. A number of scientists studied Sai Baba closely and wrote books about their experiences with him and their assessment of his psychic powers. They included Professor Erlendur Haraldsson,[3] and Howard Murphet.[4]

Sathya Sai Baba stood out in the bright orange robes he always wore, and his striking jet black Afro hairstyle. There was no mistaking him. Meeting him in person was an unforgettable experience. When he performed his miracles he would reach up and pluck an item out of the air so that sceptics could not accuse

2 Dabholkar H. *Shri Sai Satcharitra: The Wonderful Life and Teachings of Shirdi Sai Baba,* Enlightenment Press (2016)
3 Haraldsson E. *Modern Miracles: The Story of Sathya Sai Baba: A Modern Day Prophet,* White Crow Books, 2013
4 Murphet, H. *Man of Miracles*, Weiser, 1977

him of 'sleight of hand' – slipping jewelery out of the sleeves of his long orange robes.

A compelling testimonial comes from a Californian visitor to Sathya Sai Baba's Ashram. The American wanted to photograph the guru but found the film in his camera had run out and that he had omitted to bring a spare roll. Sai Baba reached up and plucked a roll of film out of thin air, suitable for the camera, and handed it to the Californian. The man took photographs and on his return to the USA he went into his local photo shop to get the roll of film developed.

When he went to collect the pictures the owner of the store remarked on the photographs. He said the man in the photos had called into his store to buy a roll of film. He remarked on it because he was an unforgettable customer dressed in bright orange robes with an Afro hairstyle. When asked when it was, the store owner recalled without difficulty. Sai Baba had walked into the shop and bought the roll of film when the Californian was standing with Sathya Sai Baba, at a loss because of his lack of film.

The remarkable fact is that apart from one trip to Kenya and Uganda in 1968, Sathya Sai Baba never left India and yet there are many reports of him appearing in countries outside of India. This ability to be in two places at the same time is called *bilocation.* Pythagoras and Apollonius of Tyanaeus were said to have bilocated. The same claims have been made for Padre Pio,[5] a famous Italian monk, who lived between 1887 and 1968. In bilocations, mystics and saints appear, in real physical bodies, in places far and from where they normally reside.

Extraordinary psychic events have been recorded since ancient times and the Bible, especially the New Testament, is an outstanding record of miraculous events. It is possible, in the new paradigm of super-energy, to account for many of these paranormal phenomena in terms of resonance between super-energy and physical-energy.

5 Allegri R. *Padre Pio: Man of Hope* Servant Books, 2000

One account of Sai Baba's materialisations, which revealed them to be *apport* phenomena and threw a light on how he performed his miracles, came from the report of an interview with Sathya Sai Baba by the famous American spiritual teacher, Ram Dass:

"When I was there, as I was sitting at his feet and he was sitting on a chair, he said to me, '"Here Ram Dass, I'll give you something." and I said, 'No Babaji, I don't want anything.' "No, no, let me give you something." He held out his hand, and I knew he did things like this, manifest small things like bracelets, watches, small things like that.

"As a social scientist, responsible to the West, my eyes were going to watch his hand closely, I wasn't going to blink. As I watched, a bluish light formed on the top of his hand, a flickering light, and it became more and more solid, and then it became a little medallion. It had a little circle with a star on it with a little gold image of himself, Sathya Sai Baba. He gave it to me, it was definitely man-made, it did not have an astral quality to it at all.

"Later I asked a Swami there, "How does he do that?" And he said, "Well, he doesn't make those; he just moves them from his warehouse with his mind. And you can just imagine his warehouse, full of these little medallions, and if you were in the warehouse, they'd be disappearing from the shelves, literally."

The report by Ram Dass helps to explain, in the *frame of super physics* how miracles performed by Sathya Sai Baba could occur as resonance between energy and super-energy. To begin with it is clear that the objects manifested by Sai Baba existed in the physical dimension prior to appearing miraculously. This is obvious from an inventory taken, after Sai Baba died, of 98kg of gold items and 307kg of silver ornaments, mainly in the form of jewellery and watches, which were found in Sai Baba's personal quarters of the ashram. Some of these may have been bought by his ashram – as in the case of the mass produced medallions – others would have been given to him over decades – mainly the expensive jewelery and watches – by his Indian devotees, as is common practice in India to give gifts to a guru.

Rather than redistribute these gifts to the poor, as the sanctimonious might have expected, Sai Baba gave them to his devotees, spiritual seekers and visitors from the West, in a most unusual way. Sathya Sai Baba lived up to his message *Love all, Serve all* by giving generously throughout his life. He gave generous aid. He built schools and hospitals and provided humanitarian support to the poor of India. But giving poor people expensive jewellery would not have done them any good.

The Westerners, who came to India to see Sai Baba, were not poor in material things but many were spiritually impoverished. They needed proof of the supernatural and spiritual dimensions to support their flagging faith and to give them firm reason to follow the four principles, which according to Sai Baba; people should live by: *Truth, Peace, Righteousness* and *Love.*

When Sai Baba came back into the world on March 8th 1940, he brought with him remarkable spiritual gifts, known in India as *siddhis,* which caused him to become one of the most famous Indian Gurus of all time. Jesus Christ had similar *siddhis*. Two thousand years before Sai Baba, the performance of healing and miracles caused Jesus to become the most renowned spiritual teacher in the West. Sai Baba likened himself to Jesus in his ability to heal and perform miracles. As with Jesus, the healing and miraculous manifestations supported the spread of the all important message of universal love. The downside for both of them was that the people thought they were 'God'. This assumption can be addressed by explaining the performance of miracles in the new super-physics frame of understanding.

According to the new paradigm of super-energy, when the soul of Sathya Narayana Raju left his body, during the coma brought on by a scorpion sting, the soul of Sai Baba of Shirdi took over Sathya's body. In New Age parlance this would be called a *walk-in*. It was Sai Baba of Shirdi who woke up from the coma in the fourteen year old body, not Sathya Narayana Raju. Keeping the boy's first name, Sai Baba continued his ministry in India that was terminated by his death in 1918. In a new body, twenty-two years after he had died, he returned with more spiritual power than he had before as *Sathya Sai Baba.*

It is conceivable to suggest that Sai Baba came from the hyperphysical world and walked into a body vacated in the physical world. He came back with all the knowledge and experience from his previous life. This would explain how the fourteen year old was able, quite unexpectedly, to quote long passages of Sanskrit philosophy and poetry and describe in vivid detail distant places unknown to Sathya Narayana Raju.

Sathya Narayana Raju probably died from the scorpion sting. Sai Baba's first actual miracle, in 1940, could have been the resurrection of Sathya Narayana Raju's body back to life; just as Jesus did when he resurrected his crucified body in the tomb.

Sai Baba, in his new incarnation, was untiring in his drive to make an impact on an incredulous generation, just as Jesus Christ did in his time. Both men used miracles to galvanise faith in people for the spiritual life and to encourage belief in the existence of higher planes of reality.

The ability of Sai Baba and Jesus to perform miracles can be explained in terms of super-energy resonance. Both of them may have been assigned hyperphysical operators of the super-energy resonance technology, designated to work with them while they were on the physical plane.

Just as the psychic surgeons worked with hyperphysical 'surgeons from the fourth dimension' so Sai Baba could have been working in association with a hyper-physical 'miracle operator' from the fourth dimension; the purpose being to renew faith in the spiritual life for millions of incredulous people.

Sai Baba certainly achieved his objective as a spiritual teacher. As a miracle worker, like Jesus, he opened the hearts and minds of millions of people – including my in-laws – reminding them that there is far more to the Universe than what we perceive in this world.

Sathya Sai Baba would have been familiar with the gifts of watches and jewellery he had been given, which he kept in his *Sai Stores.* By visualising one of these objects, during an interview with a Western seeker, Sai Baba could have thereby informed his hyperphysical operator, by telepathy, of the item he wanted. The operator, aware of Sai Baba's thoughts, would have then directed a beam of intense super-energy onto the intended object, on a shelf inside the Sai Store.

Once inside the super-energy of the beam, the energy in the atoms of the gift item would have undergone resonance to become super-energy causing it to vanish. Super-energy resonance was alluded to in the Star Trek program, but instead of receiving the order "beam me up Scotty," the hyperphysical Scotty equivalent would have perceived Sai Baba's thought as the equivalent of, "beam me up such a watch or such an item of jewellery."

As Sai Baba stretched out his hand, the watch or piece of jewellery, now in hyperphysical space, would have been beamed by the operator back into physical space just above Sai Baba's outstretched hand. Then the hyperphysical operator could switch off the beam, which would allow the energy in all the vortices of matter, in the piece of jewellery, to revert back to the speed of light. That would cause the item to reappear in physical space and drop neatly onto Sai Baba's hand.

The item of jewellery or watch, that a moment before had been in Sai Baba's store, would now be in his hand. This is because it would have been moved through *hyperphysical space* not physical space. By moving the item out of physical space and then back into it, it could appear out of thin air, ready to be presented by Sai Baba to his amazed guest.

As the super-energy in the beam, above Sai Baba's hand, descended momentarily into physical energy, it appeared as an energetic light. That was what Ram Dass saw as a flickering blue light when his medallion appeared.

Erlendur Haraldsson once asked Sathya Sai Baba why Sai Baba could perform miracles and not he. Sai Baba replied that we are all like matches – the difference is that he is on fire.

Bilocation can be understood in terms of *duplicate body replication* achieved by streaming super-energy through a hyperphysical body template. Condensation into physical matter by *descension super-energy resonance* would enable a replica living physical body to appear in another place. This process explains another extraordinary psychic phenomenon called *ectoplasm*.

CHAPTER 3
ECTOPLASM APPARITIONS

In her autobiography, *Voices In My Ear,*[1] Doris Stokes, gave a graphic description of an ectoplasm apparition:

"Silence fell, Helen Duncan concentrated deeply and then appeared to go into a trance. This was quite routine and by now I had seen it happen several times, yet there was something electric in the air. Something strange and tense that I'd never noticed before.

"As we watched a thin silvery mist began to creep from the medium's nostrils and her middle, yet she remained motionless in her chair as if she were asleep.

"Ectoplasm" someone whispered behind me. Gradually the flow increased, until the mist was pouring from the medium and a wispy cloud hung in the air in front of her. Then like fog stirred by a gentle breeze, it began to change shape, flowing and swirling, building up in places, melting away in places.

"Before our eyes the outline of a woman was being carved in mist. Hair and features began to sharpen and refine. A small nose built up on the face, then a high brow, lips and chin, until finally the swirling stopped and she stood before us, a perfect likeness of a young girl in silvery white - and she was beautiful.

"My mouth dropped open and I couldn't tear my eyes from this vision. I was seeing it, yet I couldn't believe it. Dimly I was aware that the woman next to me had gasped and clasped her hands to her mouth, but before I could register the significance of this the girl began to move.

"The audience watched, riveted as she drifted across the room and stopped right in front of my neighbour.

1 Stokes D. F. *Voices In My Ear,* Futura Publications, 1980

"I've come to talk to you mother," said the medium in a light, pretty voice quite different from the one she'd used earlier. The girl spoke to her mother, for several minutes, explaining that she still visited the family and knew what was going on and listed a few personal details as proof.

"Then unexpectedly, she turned to me. "Would you like to touch my hand?" she asked.

"Dumbly I brushed the slim, pale fingers held out to me, and then in astonishment took the whole hand. It was warm! I don't know what I had expected. Something damp, cold and unsubstantial I suppose - but this was incredible. I'd touched a warm, living hand.

"Suspiciously I glanced at the medium but she was still slumped in her chair. It was impossible. It must be a fake and yet how could she have done it? Nonplussed I sank back and stared at the girl, quite speechless.

"She smiled as if she could read my thoughts, then she raised her arm and out of the air, a rose appeared in her fingers. Gently she placed it on my neighbour's lap.

"Happy Christmas mother" she said then slowly moved back and began to shrink, getting smaller and smaller, fainter and fainter until she disappeared through the floor.

"No-one stirred. We all sat motionless as if hypnotized. The only sound was the woman next to me quietly sobbing. In her hand a deep red rose, still beaded with dew - in December. It was only later I discovered that Helen Duncan was one of the greatest materialisation mediums who ever lived."

In a cloud of *ectoplasm* (also known as *ghost-mist*) a young woman condensed into physicality. She was solid and warm, a real live person. She then melted away like early morning mist. The living body that appeared in ectoplasm was recognised, as a daughter who had recently died, by her mother in the audience.

The fact that Doris felt the hand and found it was solid and warm indicated she was not having a vision as she had of Tom, when she was a girl. Everyone else in the room saw the young woman too, so obviously physical light was reflecting off her. Doris felt her with her physical fingers so it was not an apparition. The girl was real. She was formed of physical matter, as are you and I. She was alive, a recognisable human who had died and was now standing in the room. The electrons in the physical atoms of her body were repelling the physical electrons in Doris's hand. It had to be that way for Doris to feel the hand, solid and warm.

The ectoplasm in this apparition could have been light emitting superphysical plasma, condensed from the hyperphysical into the physical level of reality. From this description it would seem to have been acting as a biofield template that enabled the condensing hyperphysical matter to differentiate into a discernable body in the physical world.

The plucking of the rose out of thin air was identical to the way Sai Baba plucked jewelry out of thin air. Maybe the operator of super-energy resonance responsible for the ectoplasm apparition also took a rose from a garden or nursery somewhere on earth, in a *super-energy ascension resonance beam* and ascended it out of physical space and then switched off the beam to *drop* it back into terrestrial space above the young woman's outstretched hand.

According to my super-physics, a super-energy deceleration technology, the reverse action of super-energy ascension resonance, could cause the intrinsic speed of energy in the subatomic vortex particles and waves in superphysical plasma to descend or condense into matter, to form the ectoplasm. Atoms of the hyperphysical matter of the girl could then have been streamed, in the super-energy resonance beam, into the condensed differentiating plasmic field of ectoplasm. In the trance Helen Duncan could have been 'between the worlds' acting as a medium supporting the flow of hyperphysical super-energy into physical space; much as Sai Baba and Jesus acted a mediums in miraculous manifestations. As the super-energy left the medium and entered the physical space of the room, it seemed to condense like mist; somewhat like steam condensing into mist as it leaves the spout of a kettle and enters a cold room.

In ancient Iran the hyperphysical operator of super-energy resonance would have been called a *genie,* from the Persian word *geni* for 'a spirit'. While only a story, the fable of *Aladdin and the*

Lamp, from Arabian Nights, is very informative. When Aladdin rubbed the lamp smoke came out of it in which a genie appeared. That was remarkably like the ectoplasm apparition Doris Stokes witnessed. The apparition of the genie in the 'mist or smoke' was similar to the condensation of the girl in the ectoplasm. The genie in the story spoke to Aladdin then began to work miracles, manifesting whatever Aladdin wanted. Was Aladdin a medium? Did he meet a hyperphysical operator of super-energy resonance who was capable of moving things – including himself – in and out of physical space.

Was the story of Aladdin and the genie purely a fable for children in ancient Persian folklore, or was it based on elements of fact? To what extent could the ancient traditions of *magic,* depicted by the story of Aladdin and other folk stories, be accounted for in terms of interactions between physical humans and hyperphysical beings? There may be a lot going on in the Universe that we have yet to comprehend.

In the ectoplasm apparition, witnessed by Doris Stokes, a medium on the physical plane supported the apparition. The young woman who appeared had obviously gone through physical death but in a hyperphysical body she appeared to be alive. That was how she was able to appear in the room.

According to the new paradigm of vortex super-physics, Helen Duncan was working with a super-energy resonance operator in the hyperphysical plane of reality. The operator then used Helen as an *anchor point* to help bring the hyperphysical girl into the physical world.

Hyperphysical matter, flowing through Helen Duncan's aura as ectoplasm, appeared to support the manifestation of the living hyperphysical body that appeared in the room. Through super-energy *descension* resonance, the intrinsic speed of super-energy could have been reduced to the speed of physical light, first in the hyperphysical matter of the ectoplasm, and then in the hyperphysical body of the girl, allowing her to take physical form in the ectoplasm so she could appear in physical space.

According to the new super-physics paradigm, *super-energy descension* resonance, the reverse action of *energy ascension* resonance, would have reduced the intrinsic speed of energy in every subatomic vortex particle in the girl's 'soul-body' from twice the speed of light to the speed of light so that she could *condense in the ectoplasm* in front of her mother, and the other people, in the physical space of the room as a living physical body.

It seems physical objects already existing in physical matter can appear and disappear by the action of super-energy resonance through the intercession of mediums like Sai Baba, without ectoplasm, as witnessed by Ram Dass and thousands of other people who visited Sai Baba. However, the Aladdin Story, Doris Stokes' experience and the appearance of ectoplasm around apparitions of ghosts as *ghost mist*, seem to suggest that bringing hyperphysical bodies into physical space requires the support of ectoplasm.

This supposition is supported by a series of ectoplasm apparitions that occurred in the recent past without the intercession of a medium. Apparitions of a spiritual Lady in ectoplasm, much like a fabled genie apparition, were seen, in recent times, by tens of thousands of people and shown to millions more on the world-wide television networks.

From 1968 to 1970, a series of Marian apparitions, in ectoplasm, occurred on the dome of a Coptic church in Zeitoun, Cairo, Egypt.[2] Night after night, for well over a year, the outline of a Lady with a halo appeared in light emitting ectoplasm which poured from her luminous human figure, and flowed over the roof and dome of the church, lighting up the night sky with its brilliance.

The ghost-like apparitions of Mary, the mother of Jesus, in ectoplasm at Zeitoun were remarkable. According to local legend, Zeitoun was the village that harboured Mary and her baby Jesus when she and her family were refugees in biblical times.

From the innumerable stories of miracles and magic that have come to us down through the ages, it seems that there have been people called *magicians* who knew certain *call-words* (spells) that could conjure up hyperphysical, fourth dimensional allies (operators) to operate magic (super-energy resonance) for them. Others appear to have been born with mediumistic powers or have come into those powers at some point in their lives. From then on they had the ability to move matter in and out of physical space with their minds.

It would seem some psychic people can manipulate, with their minds, the virtual reality we live in which, under the *spell of materialism*, we believe to be solid and substantial. Then there are illusionists who have learnt to trick our perception of reality. It is hardly surprising that they who are leading illusionists, should lead the denunciation of all paranormal phenomena as illusion.

The approach in the new paradigm is to sift through stories of magic and mediumship, and consider those that aren't easy to explain away as sensory deception – such as the apparitions of Zeitoun and the miracles of Sai Baba – as possible evidence of the existence of super-energy. Stories from ancient times, like Aladdin from Arabian Nights, help to fill the picture and we can learn from them. They help us appreciate that when super-energy beings condense into physicality, they do so by

2 http://www.zeitun-eg.org/zeitoun1.htm

anchoring to a physical object. It seems that a physical object, place or person, on the physical plane, is helpful as an *anchor point in super-energy descension resonance.*

In the Aladdin story there were two genii, one anchored on a ring and the other on a lamp. In Zeitoun, Mary anchored on the roof of the Coptic church. Once they appeared through their anchor points then the apparitions seemed to be able to move about quite freely. The report by Doris Stokes suggests, in the ectoplasm apparition of the girl, the medium Helen Duncan was used as an *apparition anchor*. Ghosts use the place they *haunt* on the physical plane as a physical anchor point.

As well as helping to explain apparitions in ectoplasm, super-energy deceleration technology could account for bilocation. Think of the extraordinary way and the speed with which duplication technology has advanced in recent decades. It is only a matter of time before we develop a technology for scanning a body, atom by atom, molecule by molecule so that the body can be digitally recorded and then be replicated from the stored digital data. Maybe that has happened already elsewhere in the Universe. Maybe a hyperphysical body can be scanned and replicated. The replica could then be inserted by super-energy descension resonance as a living three dimensional body into physical space anywhere in the world. Such *holographic inserts*, as they are sometimes called, could have enabled mystics like Jesus and Padre Pio, Sai Baba, Apollonius and Pythagoras to appear in distant places without leaving their place of domicile. By their intent, these mystics could have established an anchor point, where they wanted to land, as a target for a super-energy descension resonance beam.

Apollonius

There are stories of Apollonius and Jesus where they didn't bilocate; they just vanished from threatening situations. According to legend Apollonius disappeared from a trial instigated by the tyrannical Roman Emperor Domitian who was determined to kill him. Apollonius then reappeared several hundred miles away. Jesus disappeared from a crowd set on stoning him to death and was then seen wandering at a distance. These events could have occurred through super-energy resonance to remove the Masters from danger by *ascending* them out of physical space. They would have then been relocated by shifting the beam to a different anchor point, in physical space, at a distance from the threat. By switching off the super-energy, in the super-energy resonance beam, at the new anchor point, their physical bodies would have reappeared in the safe situation. Apollonius and Jesus could have notified the hyperphysical resonance operators *telepathically* of their desired relocation.

There are extraordinary events occurring in the world which are not understood by the scientific community and are deified by the religious community. In reaction, sceptics demonise the paranormal incidents so reliable research into psychic and supernatural phenomena, as new frontier science, is virtually non-existent in the West. Fortunately, a theoretical framework is in place, based on the vortex super-physics, to rationalise supernatural phenomena, including ectoplasm apparitions. The prejudice against the supernatural in science should therefore be reconsidered.

Until the prejudice shifts beware of accessing information on the paranormal from the internet, especially from the Wikipedia online encyclopedia. Hoaxers are active on the Wikipedia site planting malicious deceptions on pages pertaining to paranormal phenomena. This is evident on the *Ectoplasm* page where Helen Duncan has been discredited by slanderous accusations and the insertion of fake pictures suggesting that she used dolls and muslin to deceive people. There is no substantial evidence to support the malicious accusations planted by hoaxers against Helen Duncan.

Sceptics rely on science as their benchmark for truth, but the science they believe in based on materiallism is dead. Science is dead if it deals only with the material world because material substance does not exist; it is an illusion of form set up by spin.

Sceptics must understand fully that everything is energy and the apparent material world is set up by energy which is a reality that has no mass. Energy is a non-material reality. We cannot be certain about what is unseen, but then we cannot trust what we see for certain either. Werner Heisenberg made that clear to us all.

The time has come to apply science to the supernatural instead of against it. We need the scientific method to lead us out of the superstitious and religious morass associated with miracles where miracle workers are mistakenly worshiped as God. People who perform miracles are not God; they are gifted humans who can use *siddhi* powers for good or for evil.

At the same time a word of caution. Anyone interested in mediumistic power should examine their intent. If they are seeking money, power or acclaim in the world there are opportunities to achieve such dreams by psychic means but there are dangers attached to seeking supernatural powers and a very dear price may be exacted. If a siddhi comes naturally, only ever use it selflessly for the welfare and spiritual upliftment of humanity, but my advice is never, ever seek a siddhi, or become a medium if any form of pact is involved.

Most people in religion and science refuse to accept evidence that defies their belief. That intransigence is a feature of people, not religion or science. There are fanatical adherents attached to science as well as religion. There are fundamentalists in science who decry paranormal phenomena as deceptions and even discount them as a malicious deception; a *hoax*. Everyone has a right to their opinions, but a serious problem does arise when those who cry hoax become hoaxers in an attempt to falsify evidence that threatens their blind faith in materialism and their certainty that it is impossible for non-material realities to exist.

CHAPTER 4
THE HOAX DECEPTION

One of the problems with research into the paranormal is that when reports of new phenomena appear in the media and begin to arouse public interest, as occurred with psychic surgery, a malicious deception called *hoaxing* occurs to discredit the psychic phenomenon. This is very often followed by the confessions of 'hoaxers' claiming to be responsible and to have been unmasked. The media then report without nuance that the entire phenomenon is a hoax. Despite the number of genuine examples that can't be disproved, the majority of people believe what they hear or read in the media and lose interest.

The familiar pattern in the campaign of disinformation against the paranormal is that initial reports of a genuine phenomenon are quickly followed by the malicious media-supported deception called hoaxing. The media focus on the hoaxing. That creates confusion which leaves the public not knowing what to believe. It is not the paranormal that is a hoax. It is the hoax that is a hoax.

Awkward facts that challenge the establishment are nearly always suppressed. Most people disbelieve things that are discredited. It is very easy to manipulate public opinion in the controlled media and destroy credibility in anything that conflicts with what the religious, scientific and political establishment prefer us to believe. Their obvious objective is to prop up the consensus systems of belief. This is why what is false appears as true and what is true appears as false. Meanwhile, the news of really important happenings remains underground. To quote Paul Hawken from his book *Blessed Unrest*:[1]

"I wrote this book primarily to discover what I don't know. Part of what I learned concerns an older quiescent history that is

1 Hawken P. *Blessed Unrest*, Viking, 2007

reemerging, what poet Gary Snyder calls the great underground, a current of humanity that dates back to the Paleolithic. Its lineage can be traced back to healers, priestesses, philosophers, monks, rabbis, poets and artists who speak for the planet, for other species, for interdependence, a life that courses under and through and around empires."

Most scientists refuse to consider for investigation any phenomenon that doesn't fit with what is currently held to be scientific. If they were true scientists they would investigate phenomena with impartiality with a view to expanding the boundaries of science. To do otherwise is to reduce science to a sham. The so-called sceptics meanwhile actively endorse hoaxing in order to uphold their worldview based on an unquestioned belief in materialism. The use of hoaxing to discredit inexplicable phenomena flies in the face of scientific integrity and impartiality.

Until the crop-circle hoaxers appeared, it was difficult to establish a serious case against hoaxing. However, the picture of 'Doug & Dave' stumbling out of the pub, not one night but practically every night in the crop growing season, to create not one but any number of enormous crop formations, in a single night, is beyond the bounds of reason. If crop formations are a hoax there must have been an army of well organised hoaxers at work on a nightly basis in the summer season to account for all the crop circles that have appeared.

The notion of an army of tipsy hoaxers cavorted through corn fields in dead of night, tramping on rope-tied boards to produce flawless patterns of a geometric magnitude never seen before on the face of the earth, beggars belief. It is beyond the bounds of credulity that hoaxers would be devoted enough to maintain the level of activity required to account for all of the crop formations reported and photographed since the mid-1980s.

Parties of people could not work the fields, night after night, in good weather and bad, without ever being detected. They could never create the vast, complex, perfect geometric formations that have appeared, year after year in the few hours of summer darkness, without ever leaving tracks. People park vehicles and make a noise, they use torches which send out lights that can be easily seen. They leave muddy footprints and make a mess. They could never get away with the decades of unremitting trespass required to make hundreds of formations always without detection; especially when angry farmers are constantly on the lookout for them.

The precision and complexity of the vast crop formations, the way they have been repeated in considerable numbers – as would be required to be scientifically validated – and the way they suddenly appear in daylight hours as well as at night, explodes the hoax hypothesis. The assertions of hoax against crop formations has unmasked hoaxing, not to be the play of pranksters but to be an ongoing ploy of malicious deception intended to discourage serious study of the paranormal.

The intervention of so called sceptics, in support of deliberate crop circle hoaxing, reveals their true colours. Any one who employs practices intended to falsify and discredit evidence of any phenomena exposes themselves, not as defenders of scientific skepticism but as supporters of deception. It is laudable, according to the scientific method, to endeavour to falsify theories but it is deplorable, according to the scientific method to attempt to falsify evidence.

Many university professors and leaders of industry, the media and the professions take a pride in being sceptical and think that to be a skeptic is to be scientific. However, professors are not worthy of their chairs if they align with sceptics involved in the falsification and suppression of evidence to maintain the generally held scientific world view based on materialism.

It is important to keep up a healthy level of skepticism but attempts to discredit or ignore evidence that challenges entrenched systems of belief does not support healthy skepticism; it perpetuates ignorance. That is the way the medieval Catholic Church behaved toward Galileo. It is not behaviour becoming of science. People who call themselves sceptics are not sceptics, in a scientific sense, if they support or indulge in the hoaxing of crop circles because hoaxing does not support science, it demeans and debases science.

CHAPTER 5
CROP FORMATIONS

If SETI (Search for Extraterrestrial Intelligence Institute) is dedicated to scientific research into extra-terrestrial intelligence, why has SETI not investigated crop formations? If extra-terrestrial intelligences exist why should they be expected to conform to our contact expectations? Inexplicable crop formations suggest the existence of extra-terrestrial intelligence. Pictograms are an obvious option for communication in the absence of a common language.

It takes only one genuine crop formation to disprove the hoax hypothesis that all crop formations are hoaxed. It only requires one genuine, inexplicable crop formation to raise the question of the origin of the remarkable unexplained pictograms that appear in crops, world wide, year after year. While there are many thousands of unexplained formations few scientists appear to have investigated them.

One intrepid scientist who has conducted credible research, over many years, into the phenomenon of crop formations is the biophysicist, Dr. W.C. Levengood. In 1994, the *Journal of Plant Physiology* published a paper by Levengood that reported unusual anatomical anomalies found only in plants taken from crop formations.[1]

When these plants were examined under the microscope, Levengood established that the formation of the plant tissue was markedly different when taken from unexplained formations compared to samples taken from hoaxed formations. In hoaxed formations plant stems are buckled or snapped as the crop has been clearly trampled, while in the unexplained formations the stalks look as though they have been carefully laid down into a bent position with never any sign of breakage.

As a scientist, Levengood scrutinised crop formations, throughout the world, for decades, and he has documented physiological changes associated with them, not only with the plants in the formations but in the soil under them.

A formation that appeared in Cherhill, Wiltshire, England, in August 1993 was investigated by Dr Levengood. His findings were remarkable and were published in *The Journal of Scientific Exploration* in 1995. [2] The journal reported:

"...the unusual discovery of a natural iron "glaze" composed of fused particles of meteoritic origin, concentrated entirely within a crop formation in England, appearing shortly after the intense Perseid meteor shower in August 1993. Abnormalities in seedling growth was also consistent with the unusual responses of seeds taken from numerous crop formations...Presence of meteoric material adhering to both soil and plant tissues, casts considerable doubt on this being an artificially prepared or "hoaxed" formation."[2]

Dr Levengood discovered microscopic fused particles of iron, unique to crop formations, much like grapeshot, in the soil under genuine formations. He found the concentration of fused

1 Levengood W.C, *Anatomical anomalies in crop formation plants*, Physiol. Plant 92, 356, 1994

2 Levengood W.C. *Semi-Molten Meteoric Iron Associated with a Crop Formation*, J.Sci. Exploration, Vol.9, No.2.pp 191-199, 1995

particles of iron, diminished in proportion to distance from the centre of the pattern. Observing these melted particles in the soil Levengood concluded that:

"...very large amounts of energy radiation must have been involved in the formations, with the intensity falling off with distance from the centre of the formation. The science points to a complex, chaotic, thermodynamic energy field of enormous intensity, with components acting unpredictably and independently.[3]

Levengood's meticulous observations are incongruent with the hoax hypothesis; especially considering the discovery of node changes in the standing plants as well as the bent plants in the unexplained formations. The evidence defies the simplistic view of hoaxers with ropes and boards. The evidence reveals an advanced thermodynamic technology has been used to bring about the changes in the plants and underlying soil in the many genuine formations investigated over the years by Professor Levengood.

To quote author, John Mitchell, from the foreword to Nick Kollerstrom's book, *Crop Circles: The Hidden Form.* [4]

"After some twenty years of crop circle research no one yet has any idea of what is going on. Every season new and better designs appear in the cornfields. They are amazingly subtle and beautiful. Nothing in the world of art today has anything like their quality...In the early days it seemed plausible that the circles were caused by freak whirlwinds or some other weather effect. That idea became impossible after 1990 when the first elaborately-designed 'pictograms' appeared. These, obviously, were products of intelligent minds. So the theorists were divided. UFO enthusiasts believed the intelligent source to be

3 Levengood W.C & Talbot N.P., *Dispersion of energies in worldwide crop formations,* Physiol. Plant 105, pp 615-624, 1999.

4 Kollerstrom N., *Crop Circles: The Hidden Form,* Wessex Books, 2002

extra-terrestrial, while most other people took the down-to-earth view that it was all a hoax.

"The 'hoax' theory implies that unknown teams of skilled and dedicated artists are secretly at work during the summer nights, stamping or raking out large scale patterns and leaving no evidence behind. That seems the only rational solution. Yet there are so many difficulties to this explanation that experienced researchers are skeptical. No one ever detects or catches sight of these supposed circle-makers, or their cars or equipment. Certain 'hotspot' fields are watched during the summer crop formation season, yet circles suddenly appear in them over night, and nothing has ever been seen or heard.

"Then there is the problem of how these large complicated patterns could possibly be completed in the few hours of summer darkness, never left unfinished and never showing any visible error. Copyists have been commissioned to make their own circles, legally, in daylight and with no time limit. But none of these has ever managed to come up with anything to match the quality of the great unclaimed masterpieces that appear spontaneously during the British summer months every year."

The scientific observations of Levengood and other genuine scientists dispel the notion that all crop formations are hoaxes. His study of formations points to complex energy fields of enormous intensity being responsible for genuine formations. In recent decades they have increased in size, complexity and dispersal in England and abroad, yet no effort has been made by the mainstream scientific community to investigate them. The simpler style of early formations could be duplicated by hoaxers, but this changed in the 1990's when increasingly complex pictograms were just impossible to be mimicked. This development, ignored by the mainstream media, coincided with the blanket dismissal of all crop circles as hoaxes by the media and the scientific community.

Science is supposed to be impartial yet crop formations expose a lack of impartiality in science. The scientific establishment now stands shoulder to shoulder with the religious establishment in terms of defying evidence in order to defend articles of faith.

Fortunately there are still a few open-minded genuine scientists amongst the minions of materialist sham scientists. They have the courage to stand up and speak out about the urgent need to investigate crop formations. When Professor Gerald Hawkins spoke on crop formations in 1997, at a meeting of the American Astronomical Society in Washington D.C., he received a standing ovation. Hawkins had been speaking about what he called the 'intellectual profile' of the unknown artists. He said, "...the mechanics of how the crop patterns are formed is a mystery but the intellectual profile behind it all has turned out to be an even greater mystery."[5]

Crop formations provide circumstantial evidence to support the new paradigm of super-physics based on the premise of super-energy existing beyond the speed of light. Crop formations appear to validate the principles of simultaneous existence and energy speed subsets. Genuine crop formations can be explained if we allow for the existence of intelligent beings with advanced technology in a world we cannot see but a world however, which permeates our own. We only ever see pictograms in the fields. We never see the intelligent beings behind them who are aware of us and our world; the beings that use our fields as canvasses for the execution of their exquisite art.

The execution and appearance of crop formations could be likened in style to the ever elusive and mysterious British graffiti artist, Banksy. Banksy prepares stencils for a planned work of art in a studio. He then ventures out under the cover of darkness to a section of a wall or building facade and, in a short space of time when nobody is watching, he sprays aerosol paint through his stencils and then he promptly disappears leaving a masterpiece of artwork on the wall to be discovered the following morning.[6]

In line with the theory of super-energy resonance, it is conceivable that artists in the hyperphysical level of reality

5 Hawkins, G., *From Euclid to Ptolemy in English Crop Circles,* Bull. Am. Astronm. Soc., 29, p 1263, 1997

6 Banksy, *Wall and Piece,* Century, 2005

might design and prepare templates for the crop formations, much like Banksy. This would be possible because of the *principle of simultaneous existence,* which allows events and occurrences in the hyperphysical domain to overlay the physical level. The principle of energy speed subsets would enable the hyperphysical artists to operate invisibly and condense super-energy into our world. If our entire physical world (third dimension) is overlaid by the hyperphysical (fourth dimension) then the fields of Wiltshire would also be part of their overall domain. Therefore, they would be able to see the crop canvass and prepare to execute a masterful image without being seen. Of course, they wouldn't leave muddy footprints; neither would they seek prior permission from the farmer.

Just like Banksy eyeing a wall, the hyperphysical graffiti artists would look at a field of rape or corn as a potential canvass rather than as private property. But unlike Banksy they wouldn't use an aerosol can of paint; instead they might use a super-energy descension resonance beam. An equivalent of the Banksy aerosol has been caught on camera (on the internet) as an orb of light dancing over a crop with a pictogram appearing in its wake.

The orb could be the end of a super-energy resonance beam projecting intense super-energy over the invisibly stencilled crop. The orb could be the sight of the super-energy, at the end of the beam, becoming visible as it drops to the speed of physical energy and passes through the visible spectrum before reaching invisible microwave frequencies. The intense microwave energy impacting the crop under the beam, for a fraction of a second, could convert moisture in the stems of the crop into steam which would plasticise them long enough so they could be molded and laid down by energy streaming over them, through an unseen overlay design.

The laying of the crop without damaging it could be explained by application of very high temperatures for extremely short intervals of time. Levengood's discoveries demonstrated that this application of thermodynamics was evidently deployed because of the microscopic fused granules of iron found within the soil following the formation of a crop pictogram. The decrease in concentration of these fused particles from the centre of the crop formation outwards is the hall mark of an energy vortex technology.

Materialism has outlived its expiry date. In the emergent super-physics of the 21st Century, evidence from crop formations could be revisited and scrutinised. If that ever occurs investigators should ask where the super-energy artists come from? Do they originate from levels of super-energy associated with the Earth or are they associated with UFOs coming from deep space? Also, what is the purpose of their art? Is it art for art's sake or is it an attempt to communicate with us and maybe broaden our minds to greater possibilities than we allow for in our current limited scientific frames of reference.

CHAPTER 6
UFOS

Since World War II there have been innumerable reports of unidentified flying objects. Books have been written on the subject and films have been made but the possibility of extraterrestrial visitors from outer space traveling through our terrestrial skies has been dismissed by mainstream scientists. In books on the UFO subject, such as *The UFO Phenomenon* by J.Von Buttlar[1] and Timothy Good's *Above Top Secret* [2] the authors claim that reports of UFO's are actively suppressed by the establishment. That makes any serious scientific study of the phenomenon difficult.

1 Von Buttlar J. *The UFO Phenomenon*, Sidgwick & Jackson, 1979

2 Good T. *Above Top Secret*, Sidgwick & Jackson, 1987

Nonetheless, thousands of reports and eye witness accounts, included in these and other books on the subject, show UFOs appearing on radars and being photographed from the ground and from aircraft. This plainly suggests that UFO's are physical objects. They must be because they can reflect visible light and also radar beams. However, numerous accounts attest to the fact that UFOs can also vanish quite suddenly and in some cases appear or reappear instantaneously. It is this last piece of information that gives us a clue as to how extraterrestrials could be traversing the vast distances between us and other planets.

Scientists and sceptics dismiss UFOs on the grounds that distances in space are just too vast to be traversed. With the enormous distances between stars and galaxies in the Universe, travelling from one planet to another through space would seem to be all but impossible. However, in the new paradigm, based on super-energy, it is proposed that if crop formations are created by other dimensional beings, using super-energy resonance, then maybe there are groups of beings on other planets in the Universe that can use that same technology to reach us. If this were true, then *super-energy resonance* could provide a link between crop formations and UFOs.

It is possible that intelligent beings could be living on planets in other star systems. Amongst the billions of stars and trillions of orbiting planets in our galaxy alone, there must be planets with conditions suitable for supporting biological life. If *super-energy resonance* is a reality, then some species, more technically advanced than us, may have mastered it as a technology for inter-stellar travel.

Although the idea of *super-energy resonance* is still just conjecture, it is a powerful idea as it explains a host of paranormal phenomena and also provides a common link between them. *Super-energy resonance* could also provide a precedence as well as a prediction for future lines of serious amounts of research that could be of great benefit to human kind.

If *super-energy resonance* is a reality, then it makes sense that somewhere in the Universe intelligent agencies would have developed a *super-energy resonance technology* for accelerating and decelerating the intrinsic speed of energy to allow for inter-dimensional travel. *Super-energy resonance* would enable a solid space craft, made of atomic matter, to move between physical third dimensional space and time and higher dimensions of space and time – along with the craft occupants. Moving in and out of physical space and time, rather than being limited to just moving through it, opens up the possibility of unrestricted travel throughout the Universe.

The possibility of unlimited travel is not only between star systems inside our own galaxy but between galaxies. This is because the *super-energy resonance technology* could allow for a space craft to exit the space-time continuum on one planet and then enter and appear in another continuum associated with another planet anywhere in the Universe.

By moving in and out of physical space and time rather than moving through it, a craft could leave physical space-time at one point and re-enter it at another, billions of miles away, almost instantaneously. This is because it would not have travelled through deep space, it would have lifted out of the local space-time continuum associated with one planet, and dropped back into the local space and time associated with another via hyperspace. As such it would not have moved relative to physical space or time. This type of inter-space travel could help travellers and explorers overcome the vast gulfs of space and immense spans of time associated with deep space. That would enable them to appear instantaneously and anywhere relative to physical space and time.

All this is possible if super-energy exists, if we are multi-dimensional beings living in a multi-dimensional Universe, and if *super-energy resonance* is a reality. If those three 'ifs' are for real then bodies of matter should be able to move, with ease, in and out of space and time. This could be achieved by alternately accelerating and decelerating the intrinsic speed of energy in every vortex and wave particle within them alternately evaporating and condensing them out and in of physical space and time. This sums up the science of UFOs and extra-terrestrial travel, according to the *super-physics paradigm* originating from the vortex theory.

It is possible that UFOs are small reconnaissance vehicles, maybe unmanned remotely controlled drones, which are dropped or 'condensed' into terrestrial space-time from mother ships or biosatellites moored in hyperspace. This could happen by *descension resonance*. If the *super-energy resonance* beam tracking and navigating them was switched to *ascension resonance* mode they would vanish or 'evaporate' out of space and time. If it were switched to *descension resonance* mode again they would reappear by 'condensing' back into physical space and time again. This may be how UFOs suddenly vanish and then reappear again.

The possibility of travel between worlds by changing the intrinsic speed of energy in matter from physical-energy into super-energy opens up new vistas for humanity with the opportunity for unlimited space travel. The possibility that super-energy resonance might overcome the limitations of space and time and open up the Universe for future exploration is hopefully prescient. This possibilitty should be sufficient justification for it to warrant investigation in science laboratories.

This research may be underway already. It would be carried out in secret quite obviously by governments or wealthy space travel pioneers like Elon Musk. Research of this nature would be likely to happen, if it is happening, outside of public scrutiny, in the interests of economic, political and military advantage.

There is also a chance that the research is not underway. Lacking impartiality and in the unrelenting grip of the materialistic paradigm, maybe the science community is not choosing to pick up this baton of research. The conformity to consensus scientific beliefs – in no way different to intransigence associated with religious beliefs – along with the scepticism prevalent in the universities, and fear of peer rebuke is likely to cause professors to use their positions of authority to oppose, rather than encourage, research into UFOs, as they do with anything else associated with the paranormal. In order to protect the tenets of materialism, the scientific establishment may obstruct the new frontiers of research into space travel. However, this may be a blessing in disguise.

Can humanity be trusted with *super-energy resonance technology*? The *Avatar* movie presented a stark warning of the threat we pose to other planets in the Universe. Were we to colonise other worlds with the same degree of selfishness, greed, irrationality and violence demonstrated within our own terrestrial sphere, then we could end up as a species as a planetary pathogen. If we behaved toward other planets as we behaved during the colonisation of Africa, America, and Australia, we could reap havoc in the Universe. We have many lessons to learn in anthropological respect, ecological management and planetary care before we are ready to take the next giant leap into space.

CHAPTER 7
THE PHILADELPHIA EXPERIMENT

In the early 1930s, a project of research into invisibility commenced at the University of Chicago. The team was headed by physicist Dr Kurtenhauer and Dr J.Hutchinson, then Dean of the University. In 1934 the project moved to Princetown under the auspices of the Institute of Advanced Study, which became involved in the Manhattan Project that developed the atomic bomb. The Institute of Advanced Study included Albert Einstein, the brilliant quantum physicist, John Von Newmann (who was largely responsible for the development of computers in the USA) and the genius researcher T. Townsend Brown.

In 1943, the US Navy gave the go-ahead for the invisibility research team to perform an experiment on a naval ship moored in Philadelphia harbour.[1] A test was run to see if it was possible to make a ship invisible to radar and visible light that could help to bring about an end to World War II. A destroyer, the USS Eldridge, was chosen for the experiment. Against the advice of the scientists involved, the Navy insisted the crew were left onboard during the experiment, which worked but disastrously.

According to eye-witness accounts the ship vanished from the harbour of Philadelphia for fifteen minutes but was reported to have re-appeared in the harbour of Norfolk Virginia, several hundred miles away before reappearing in the Philadelphia harbour. The experience was catastrophic for the crew. Many of the sailors went out of their minds and five of them were partially reconfigured into the atomic structure of the ship. Their hands were merged into the steel bulkhead, where they had been touching the metal at the time of the

1 Berlitz C. *The Philadelphia Experiment*, Souvenir Press, 1979

experiment, and had to be amputated. Needless to say the project was abandoned and classified as Top Secret.

The USS Eldridge

After the war, the invisibility project was renewed under the direction of Dr. Von Newmann at the Brookhaven National Laboratory until Congress disbanded it in 1967. In *The Montauk Project,*[2] author Preston Nichols claimed that the invisibility experiments were continued from 1971 as a secret military project under the continued direction of John Von Newmann, at the Montauk Air force Base on Long Island, New York.

In his book, Nichols claimed that vortex teleportation beams were constructed at Montauk, which enabled projections in and out of space and time. He claimed that an inter-dimensional link between the Philadelphia experiment in 1943 and the Montauk experiment in 1983 caused parts of the ship and two crewmembers to be teleported from the SS Eldridge in 1943 to the Montauk base in 1983. This story was the theme of the movie *The Philadelphia Experiment.*[3]

The Philadelphia experiment and the Montauk project both suggest that super-energy resonance technology has already been achieved and used on the surface of the Earth for shifting physical bodies in and out of physical space-time. It could be that during the Philadelphia experiment, and subsequent Montauk experiments, the intrinsic speed of energy (the Einstein constant of relativity) was altered inside every particle of energy being subjected to the

2 Nichols B. P. *The Montauk Project*, Sky Books, 1992

3 Ziller P. (Director) *The Philadelphia Experiment*, DVD

experiments. That might have caused a dimensional shift to occur as a result of physical-energy ascending into super-energy and then descending back into physical-energy again. If this actually occurred, then the implications are staggering.

The fact that the crew on the Eldridge, in the Philadelphia experiment, survived the ordeal makes it clear there was no increase in frequency of vibration. That would have caused a temperature rise, which could have killed them.

If, in the *Philadelphia Invisibility Experiment,* the S.S. Eldridge and its crew went 'beyond the speed of light', then scientists may have stumbled on the super-energy resonance process, a process conceivably used by advanced galactic intelligences to operate their UFO space crafts in and out of the space-time continuum. If the Philadelphia Experiment story is true then these scientists would seem to have demonstrated that it is possible to transfer a craft and its occupants, from physical space and time into another dimension of space-time and bring it back again.

While disastrous things happened to the crew, out and about on the ship, according to the reports nothing untoward happened to the technicians in the control room inside the ship who were operating inside the Helmholtz vortex coils that caused the ship to vanish. This demonstrated that the super-energy resonance process used had the potential to be safe for people if it is properly researched and developed in a responsible and controlled manner. In the subsequent research at Montauk, the safety issues were resolved and, according to Preston Nichols, writing in *The Montauk Project,* test subjects continued to be successfully projected in and out of time as well as space. In his follow on books he detailed how subjects came back from 'forward time' with fascinating predictions for the future.

Under the weight of the current scientific orthodoxy, open research into super-energy resonance is unlikely. Nonetheless super-energy resonance research in the controlled environment of a university laboratory, along the lines of the experiments conducted at Philadelphia and Montauk, could provide scientists with a monumental opportunity to advance their understanding of physics. It may be possible that in our lifetime we will have an opportunity to travel in and out of space-time, with other potential benefits including the possibility of reversing the ageing process and maybe even achieving physical immortality.

BOOK VII

THE PROPHECY

Life just keeps on going, free of feelings of suffering, joy, sadness, pity; like a river it tumbles each day into the next. How you wake up and perceive the day is up to you. How you choose to perceive others and yourself is also up to you. But notice how judgment and anger leaves you feeling inside; notice how when you don't communicate with people you are left feeling suffocated, blocked - even sick. Is the feeling of being right or better than others worth your own suffering? Notice then how when you give with love and understanding you are left with the feeling of joy and wholeness. There's no right or wrong, no good or evil, only a fresh new day with every sunrise. Finish each day knowing that everything is complete; there is nothing more to do or worry about. Relax into sleep in the knowing that there is nothing more important than the day that will awake you.

Rebecca Ash, 2004

ALL LOVE EXCELLING

They say you will be waiting for me when I arrive
That it will be just as it was before, and more
They say you will greet me in an ethereal gown
And we will communicate most wondrously
Moreso than ever we knew
And there will be other of family, stretching on and on
Across myriads of cosmic miles
Through oceans of plasmic space
No longer measured by earthly time
Back and back generations
Onwards and towards the Divine
Friends will embrace us, and angels will sing
As they help lead the way, to the all seeing and all being
With no bags, no belongings, or so it would seem
For the mind and the body is all spirit beyond dreams
All this, all I know now, in this home, on this Earth
Which has housed me so well, since the day of my birth
Will one day be replaced, when it's time to move on
Beyond the fine line which divides us
From the spiritual realm
All love excelling. All blessings from above
Deeply I thank you with all of my love
Oh Gloria in the Highest. Look well on us here
As we travel our earthly life
HALLELUJAH

Susan Saillard-Thompson © January 2016

CHAPTER 1
ASCENSION

In 1991, on a visit to Australia, I was introduced to a set of five cassette tape recordings originating from America that predicted an event called: *The Ascension.* An American, by the name of Eric Klein, had made these extremely compelling voice recordings and said that he channeled them from the fifth dimension. Subsequently published in his book, *The Crystal Stair,*[1] the channellings detailed how the predicted ascension would occur and how humanity could prepare for it.

I was galvanised by the ascension information. As a vortex physicist I was aware that the process of *super-energy resonance* provided a viable scientific underpinning to this phenomenon. I was electrified by the compatibility of this spiritual transmission with the vortex model in physics I had been developing for twenty years prior.[2] Subsequently, I devoted four years to lecture tours around the world propagating a fusion of paradigm busting physics and transformative spirituality based on the ascension prediction.

The message from the super-physical, fifth dimensional intelligence – speaking through the physical conduit, Eric Klein, in our third dimension – was that humanity is living through a singularly unique period of planetary history. Everyone on the Earth is going through an immense spiritual transition. The message also spoke of how our planet is approaching a time of rapidly sweeping Earth changes. In the Eric Klein channellings it was explained that there would be three waves of ascension giving everyone an opportunity to ascend from the third dimension to the fifth dimension. The waves would coincide with three increasingly disruptive events on the Earth. The third

1 Klein E. *The Crystal Stair,* Oughten House, 1992
2 Ash D. & Hewitt P. *The Vortex: Key to Future Science,* Gateway Books, 1990

wave, just before the final Earth changes, would include an evacuation of babies, little children and innocents. An assurance was given for those who choose to ascend that children and pets in their care will be evacuated with them.

The main superphysical source of the Eric Klein channellings, known in the Vedic scriptures as *Sananda* and in the Christian scriptures as *Jesus*, described the ascension process as follows:[3]

"So what is the process? You have grown accustomed to reincarnational experience – that is, having your soul incarnate in a body and going through a lifetime of a certain number of years, experiencing death, and leaving your body to return again to the earth. You have gone to some fourth-dimensional areas. There are heavens and hells galore, with many experiences. Yet always you return to the body, for the body is the platform from which your launch will take place into the fifth dimension. This is the intention, let me say, of having a physical body, at this time especially. I would say that your ascension will be all but identical to my own. You will not leave your bodies behind and go to a higher state of consciousness. Your bodies will be transformed also. The molecules and atoms, your subatomic particles, all that you are, will be transformed and accelerated into the fifth dimension. So you do not have to die. Well, that's some good news. Despite the fact that human beings have grown so accustomed to dying that it has become a common awareness or common belief, I am telling you now: you do not have to die. And you will not."

In the Eric Klein voice recordings and in his follow on book it was predicted that human beings will be offered the opportunity for evacuation and then ascension, prior to the Earth changes, through the experience of a phenomenon of light. A doorway of light will appear, without warning. Each of us will have the same opportunity to enter into it. Whether awake or asleep, we will all be fully conscious of the door of light. The door of light will last only a few moments. If we hesitate or choose not to enter, it will fade away.

3 Klein E. *The Crystal Stair*, Oughten House, 1992

In *Emissary of Light*[4] James Twyman wrote of his experience of the door of light and described it as the 'Door to Eternity'. He explained that we will be drawn into the light of the 'Door to Eternity' by the light in our hearts. We have nothing to fear and nothing to worry about. All we have to do is to let go of our judgments of ourselves and others, our apprehension and our reservations and trusting the light, surrender to it when it is offered to us. The ascension prediction fits with the *rapture* predicted by Christians, when the elect are *lifted to heaven* to save them from the *end time events*.

In the Eric Klein channellings it was made clear that *the elect* are not an *elected few* but are *the few who elect* to enter the door of light when it is offered to them. If the ascension predictions are true, entry into Heaven (the fifth dimension) would seem to be a *self-selection process*. No heavenly agency is judging us. We, each of us, will judge ourselves.

Ascension Door of Light

Civilisation with all its distractions, distortions and temptations could be acting as an *Ascension filter*. Never in history have there been so many things available for people to become attached to on the physical plane. In the final days of this age the population may be about to 'sit their spiritual finals'. How will we fare in the spiritual examination which is essentially a test of our priority for spirituality over materialism? Will we be able to let go of the physical for the sake of the spiritual? In a few moments of decision, when the invitation to ascension opens before us as a door of light, will we be able to let go of everything material we are attached to in the physical world in order to step voluntarily into the spiritual world? This is the choice point?

4 Twyman J. *Emissary of Light,* Warner Books 1996

The vast majority of people who ever lived have died. Only a tiny minority have ascended. At death we have to leave everything behind, but death is involuntary. We have no choice. Ascension is voluntary. In order to ascend we have to choose to leave the physical plane and leave everyone and everything physical behind. In that respect, it may be an exact equivalent of death, but the criterion is different. With death there is no choice. With Ascension we have to choose.

According to the ascension channellings, those who ascend in the first wave will be offered an opportunity to return to Earth in the ascended state to act as way showers for the rest of humanity. Those who return after the first wave of ascension will be able to help people prepare for the second and third waves of ascension. They will also help to alleviate suffering during the unfolding end-time events.

Those who do not elect to ascend will die sooner or later. Death may come shortly after the 'event of light' if the changes to the face of the Earth proceed as predicted by the earth sciences (detailed in the next chapter). How death comes is inconsequential. It will come eventually to all who don't ascend.

In the Eric Klein channellings it was said that souls can ascend from the fourth dimension (hyperphysical plane), where people normally go after death so no one will be lost if they are ready to ascend. And there is no doubt the Earth could benefit from purification before we completely destroy it. In that sense everything we may choose as priority over ascension will go anyway, when civilisation is swept away – as has happened to every civilisation that has gone before our own.

Super-energy resonance provides a scientific account for how the ascension could happen; if it happens. The opening of light, acting as an invitation to ascension, could be the end of a fifth dimensional *super-energy resonance beam.* If so, as each one of us steps into this doorway of light, or even moves toward it in our minds, we would be choosing to enter a beam of superphysical super-energy. Should we decide to do so every single particle of energy in our body would resonate with the super-energy contained inside that beam. During this resonance, between the worlds, the intrinsic speed of energy in the physical body would accelerate to a point of equivalence with the speed of the super-energy in the beam. We would then experience a 'lifting' and find ourselves 'beamed' out of the physical plane and into a safe-holding in a higher dimension. Like Captain Kirk, we would have been 'beamed up' from what is likely to become 'a hostile planet' if the predicted Earth changes proceed.

I trusted the channellings because of who they came from. While I was no longer a practicing Christian nor did I believe Jesus is God, I believed and still do, that he was one of the most authentic men in history and I have implicit faith in his authenticity as a planetary guardian. When the voice that came through in the first channeling, identified as Jesus, my heart told me it was him. I trusted then that what was said on the tapes was true.

I was also excited by the idea of ascension because of the vortex super-physics. I realised immediately that *super-energy resonance* could account for the ascension. I imagined ascension could happen through the same technology I had predicted to account for miracles performed by Jesus when he was on Earth. I perceived it as the same science behind UFOs, psychic surgery, Marian apparitions and Sai Baba's manifestations. I had referred to all of this in my book *The Vortex: Key to Future Science* [5] prior to hearing of the ascension.

I had no doubt that the ascension, if it occurs, would be an extraordinary gift of love and mercy from the Godhead to humanity. The use of the word Godhead I define here as an immensely focused manifestation of sentient superphysical energy. Due to the universal and all inclusive nature of this level of being there is no judgment or discrimination on the part of the super-beings, from that level of compassion and universal love, they want to help each one of us.

I sum up the Eric Klein channellings on ascension in one word: compassion. Out of compassion everyone will be offered an equal opportunity to take a leap of faith and enter a doorway

5 Ash D. & Hewitt P. *The Vortex: Key to Future Science*, Gateway Books, 1990

of light and so avoid having to go through the trauma of Earth changes. However, the decision made to step into a vortex resonance teleportation beam takes courage and crucially rests on a person's capacity for faith and surrender to a higher plan, which largely depends on their commitment to spipritual values over many lifetimes.

It is difficult to prepare for ascension as no one knows for sure if or when it will happen and ascension has nothing to do with good or bad. There is no judgment in the ascension process. Ascension is all about total acceptance of everyone and everything. It is a manifestation of unconditional love. In ascension we do not lose anything, we gain everything.

The evacuation from the Earth could be likened to people who are asked to leave their home temporarily not knowing it is going to undergo an *Extreme Makeover* as a total surprise to them. Their subsequent transformation into the 'ascended state' could be likened to a pauper living in a cottage who is discovered to be heir to the throne. When the king dies the pauper comes into the kingdom and moves into the palace instead of the cottage. The cottage is still his so he can visit it anytime because it belongs to his kingdom. In like manner after ascension we can return to the Earth anytime we like.

Ascension and the Earth Changes are not the 'End of the World'. They are the beginning of a new Golden Age of love and light after the end of the dark age of Kali Yuga. These changes for us and the Earth are reason to celebrate, not grieve.

Our decision, whether or not to go for evacuation and ascension will be taken on the basis of how we have lived our lives and the choices we have made, every day, between love and fear. This will be the major influence on our decision during the pivotal moment of greatest opportunity imaginable. How we choose to live, day by day, determines not only our destiny in this world but also in the worlds to come. If we live out of love rather than fear, surrendering criticism and judgment, of others and also of ourselves, if we manage to overcome the addiction of attachment to people, places and things then we will be better primed to let go and enter the opening of light to be a beneficiary of this planet's ascension process. This is not a call to give up our responsibilities, our jobs or our homes. We need to be 'fully grounded' to fulfil our daily commitments and live life to the full, but at the same time, have the ability to release our attachment to the things of this world. As Ry Cooder sang, in *The Prodigal Son*:

If you'd like to get to heaven and see eternity unfold,
You must, you must unload. [6]

Most important, in the ascension process is the realisation that nothing matters apart from love. Ultimately it doesn't matter if we ascend or not because it will happen for us all, eventually, when we are ready for it. It is not good to become obsessed with ascension as that can block the ascension process. An attitude of fearlessness and non attachment to ascension – as well as everything else that pertains to the mind – is necessary in order to pass spontaneously into the light. Ascension is all about the ability to take a leap of faith, to live in the moment, to be in the flow rather than being stuck. Ascension happens in an instant of spontaneous abandonment to the heart rather than slavishly following the mind. Ascension occurs in the eternal 'now' so if we practice unconditional love, learn to let go and live life as it happens rather than how we think it should happen, then in the moment we are called ascension will happen for us as easily as moisture evaporating off blades of grass when the sun rises.

My belief in the imminence of ascension is reinforced by the *Earth's Shifting Crust,* [7] a seminal book written in 1958 by Earth scientist, Charles Hapgood, and endorsed by Albert Einstein.

6 Cooder R., *The Prodigal Son* Track. 7, You Must Unload, Perro Verde Recordings LLC, 2018

7 Hapgood, C., *Earth's Shifting Crust: A Key to Some Basic Problems of Earth Science,* Pantheon Books, 1958.

CHAPTER 2
THE EARTH CHANGES

In the *Earth's Shifting Crust* [1] Earth Scientist Charles Hapgood presented compelling evidence that the crust of the Earth periodically moves over the main body of the planet. Hapgood pointed out that major Earth changes occur cyclically in a period which is less than ten thousand years.

In his book of seismic implications, the author rewinds the earth clock some ten thousand years. The geological evidence shows that rivers were flowing on the continent of Antarctica less than ten thousand years ago while the northern States of America were covered under a sheet of ice miles deep. The pole was at what is now Hudson Bay.

Hapgood presents disturbing evidence that less than ten thousand years ago something happened that caused America to be displaced some two thousand miles towards the equator, shifting the pole from Hudson Bay to its current position in the Arctic Ocean. It happened so suddenly and so swiftly that Mammoths in Siberia were flash frozen with grass still in their mouths.

According to NASA, after inching back toward Hudson Bay, since 2000 the North Pole has been heading South toward Britain at the rate of 17 cm per year. According to Hapgood's hypothesis this would be caused by Earth's crust sliding in the opposite direction over the Pole, moving Europe North.

Mainstream Earth sciences reveal that every few thousand years disparate regions of the Earth are visited by radically different climatic conditions from polar ice to desert sands, from equatorial rainforest to temperate woodland. It's difficult to believe that the

1 Hapgood, C., *Earth's Shifting Crust: A Key to Some Basic Problems of Earth Science,* (foreword by Albert Einstein) Pantheon Books, 1958.

Lake District of Northern England was once covered by a glacier, while just a few hundred miles to the south the evidence is that the red sand cliffs of East Devon were once desert dunes. Travelling along to the Jurassic coast in the neighbouring county of Dorset we find evidence of dinosaur bones. The implication of this is staggering. These cold blooded reptiles lived in hot climates. They could only survive in tropical and equatorial regions of the Earth. All the evidence of the Earth sciences points to dramatic climatic changes periodically occurring all over the Earth, which could be explained by the crust shift hypothesis.

If the entire crust of the planet were to slide over the poles the resultant catastrophe for human life is unimaginable. Let's put this in perspective. The Atlantic Ocean is about twelve thousand miles long and two to four thousand miles across and on average two miles deep. On that scale the oceans on the planet are the equivalent to a millimeter deep spill of water on a table top. Now think of human civilisation on this scale. Even the tallest buildings and greatest cities wouldn't even register as a miniscule film on the surface of the Earth.

There is no way any civilisation could survive a cataclysm in the magnitude of a roll of the entire crust over the poles of the Earth. Practically everything and everyone would be destroyed by the resultant earthquake and fire, storm and flood. Only a tiny remnant of the human race would survive.

If a disaster on this scale happens to our planet, every few thousand years, then obviously we are living on a very unstable planet and we are in a very precarious position. That is why the message of ascension is so important to us all and why materialism is such a sinister trap.

Earth sciences show us that pole shifts have occurred repeatedly in the course of the Earth's history. As we move in time further from the last pole shift we inch ever closer to the next one coming. That is why prophets and messiahs, spiritual teachers and enlightened scientists keep warning us to beware of materialism and to be prepared for a day of reckoning that is coming; a day when everything 'human' will be wiped off the face of the Earth and the remnant survivors are plunged into a new Stone Age.

We don't know when the next pole shift is due. I personally believe it could be any day now because, in my book, pole shifts

are linked to global warning. My account begins with the man of the 20th Century, the most enlightened scientist in modern times; Albert Einstein.

Albert Einstein was immediately impressed when he read Hapgood's manuscript and he wrote a foreword to the book.[2]

In Einstein's words:

"I frequently receive communications from people who wish to consult me concerning their unpublished ideas. It goes without saying that these ideas are very seldom possessed of scientific validity. The very first communication, however, that I received from Mr. Hapgood electrified me. His idea is original, of great simplicity, and – if it continues to prove itself – of great importance to everything that is related to the history of the earth's surface.

"A great many empirical data indicate that at each point on the earth's surface that has been carefully studied, many climatic changes have taken place quite suddenly. This, according to Hapgood, is explicable if the virtually rigid outer crust of the

2 Hapgood, C., *Earth's Shifting Crust: A Key to Some Basic Problems of Earth Science,* (foreword by Albert Einstein) Pantheon Books, 1958.

earth undergoes from time to time, extensive displacement over the viscous, plastic, possibly fluid inner layers. Such displacements may take place as the consequence of comparatively slight forces exerted on the crust, derived from the earth's momentum of rotation, which in turn will tend to alter the axis of rotation of the earth's crust.

"In a polar region there is continual deposition of ice, which is not symmetrically distributed about the pole. The earth's rotation acts on these unsymmetrically deposited masses, and produces centrifugal momentum that is transmitted to the rigid crust of the earth. The constantly increasing centrifugal momentum produced in this way will, when it has reached a certain point, produce a movement of the earth's crust over the rest of the earth's body, and this will displace the polar regions toward the equator.

"Without a doubt the earth's crust is strong enough not to give way proportionately as the ice is deposited. The only doubtful assumption is that the earth's crust can be moved easily over the inner layers..."

Albert Einstein pointed out that for Hapgood's account to be plausible, an explanation had to be found for heat in the Earth sufficient to enable such massive periodic events. For the crust to slip the heat in the Earth would have to be sufficient to turn the *magma* (molten rock) under the crust from the consistency of glue to that of a lubricant (a bit like turning sticky honey to runny honey when it is heated) so as to enable the crust to slide freely over the inner layers.

The Hapgood hypothesis requires not only that there is an enormous source of heat in the Earth but that the colossus heat is released periodically. Something extraordinary would have to be going on in the Earth to cause regular Earth changes of such a magnitude as to enable the crust to roll entire, over the poles, not just once but again and again.

In *Earth's Shifting Crust*, Hapgood admits that the heat in the Earth is a mystery. In fact, according to experts it is one of the main unsolved problems in geology. He quotes from *The Internal Constitution of the Earth* by Beno Gutenberg:[3]

"A vast amount of research has been devoted to this subject, but the fact remains that the origin and maintenance of the earth's internal heat continues to be one of the outstanding unsolved problems of science."

The theory for gravity, in *The Vortex Cosmology*, provides an account for heat rising periodically from the core of the Earth, sufficient to cause periodic pole shifts in line with Hapgood's hypothesis. Furthermore, global warming may be linked to the rise of heat from the core to the crust of the Earth. The vortex account provides for the heat to be sufficient to reduce the viscosity of the molten rock under the crust so that instead of acting as glue, gripping the crust to the inner layers, it acts as a lubricant enabling the crust to slide freely over the inner layers of the Earth.

As I developed the vortex theory, I realised I could account for the mystery of the heat in the Earth. I discovered that a vast amount of heat could be generated from the *annihilation of matter and antimatter* at the Earth's core centre.

Matter-antimatter annihilation is the greatest known source of energy in the Universe. Expressed in Einstein's equation $E=mc^2$, it incurs 100% conversion of the mass of subatomic particles into energy; quite sufficient to turn molten rock (magma) from the consistency of sticky glue to that of slippery oil. This account for the heat generated by the Earth is only available in the vortex theory. Not only does it explain how magma beneath the crust of the Earth can turn into a lubricant, it also implies that this heat could account for global warming. My thesis suggests as we are in a phase of global warming the crust of the Earth could shift at any time.

Because of my understanding that global warming could be the advent of a shift of the Earth's crust , I take the ascension message very seriously. I feel a profound sense of urgency to

3 Gutenberg B., Internal Constitution of the Earth, Dover, 1951

warn as many people as I can of what might be happening right under our feet. This is not to spread fear but to encourage us all to rethink our priorities; especially with regard to the spell of materialism and the attachment it brings to the things of the Earth.

If pole shifts are inevitable and occur without warning, during periods of global warming, a slip of the Earth's crust could be imminent. That would mean everything we perceive as valuable by today's materialistic standards such as money, and property, the internet, fast-cars, careers, investments, Picasso paintings and such like, could disappear in a day. This is because no civilisation could survive an event of such a magnitude as the crust of the Earth sliding over the poles.

The legends of lost civilisation like Atlantis, Lemuria and Mu could be based on fact. Stories, epic poems and ancient records from history tell of great civilisations disappearing, without a trace, as a result of periodic global catastrophes. Most people dismiss these legends. I do not.

Using the vortex theory I am able to explain not only how enormous amounts of heat could rise from the centre of the Earth, due to the annihilation of matter and anti-matter, but how this would be periodic. Latching onto Stephen Hawking's idea of *sub stellar mass black holes,* [4] I propose a possibility that there might be a sub-stellar black hole at the centre of the Earth. Using this proposition I can account for how the release of vast amounts of annihilation energy from the centre of the Earth would be periodic and why crust slips would occur like clockwork in the Earth's history.

In *The Vortex Theory* [5] I suggested that antimatter exists in equal amount to matter beyond the centre of the Earth. Annihilation could occur where matter meets antimatter in the

4 Hawking S. *Gravitational collapsed objects of very low mass,* Monthly Notices of the Royal Astronomical Society, 1971

5 Ash D. *The Vortex Theory,* Kima Global Publishing, 2015

black hole in the centre of the Earth. While the core of the Earth would not be pulled into the black hole, because of its atomic structure, energy of annihilation would rise from it.

If the internal heat in the Earth originates from matter-antimatter annihilation, it would be about two hundred times greater than that derived from nuclear fusion in the sun or the hydrogen bomb. If my vortex theory is correct, there could be vast amounts of heat coming up from the core centre of the Earth sufficient to turn sticky molten rock into a lubricant liquid.

But why would the crust slip be periodic? Why would a displacement of the Earth's crust be likely to happen in our day, coincident with a period of global warming? In developing my theory it struck me that the periodicity could be caused by the way black holes behave. In *The Vortex Theory* I explained how the attraction between matter and antimatter, through the centre of black holes, causes extreme gravity. Then in my account for gamma ray bursts from distant galaxies I postulated that black holes behave as *geysers*. I described black holes as *gravity geysers*. The gravity geyser hypothesis I developed to explain black hole behaviour in distant galaxies I then applied to the Earth.

If the heat in the Earth were coming from the annihilation of matter and antimatter, energy would be generated continuously in the sub-stellar black hole at the core centre of the Earth. However, the energy would be gripped initially by the intense gravity of the black hole. Then, when the build-up of annihilation energy is sufficient to overcome the ability of the gravity of the black hole to hold it, some of the trapped energy would escape to rise through the dense core of the Earth. The release of heat would not be immediate because the gamma ray energy produced by the annihilation would be first transferred into heat – by bouncing around in the dense core of the Earth – only then would a wave of heat from the energy release be able to travel thousands of miles through the inner layers of the Earth. It would reach the layers of molten rock magma under the crust, through convection.

After a release of energy, the gravity of the mini black hole at the centre of the Earth would 'snap back tight' on the annihilation energy. The core would then cool down ready to start the cycle all over again. Meanwhile a wave of heat would be rising, slowly yet inexorably, toward the crust of the Earth, destined to destroy any human civilisation residing upon it.

The heat originating from annihilation between matter and antimatter, could be the missing link in Hapgood's theory of the *Earth's Shifting Crust*. A *gravity geyser* at the core centre of the Earth could be the missing piece of the Earth Science 'jigsaw puzzle'. It is speculation but it could explain why periodic waves of heat rise again and again from the heart of the Earth. The release of annihilation energy by a gravity geyser could explain the planet-wide catastrophes that seem to happen periodically wiping out civilisations without trace and why this keeps happening. The gravity geyser theory accounts for climate changes, periodic pole shifts, mountain formations and periods of global warming followed by ice ages that constantly occur on our planet

The regular cycle of the release of immense heat from the Earth's core causing the Earth's crust to float would not in itself be the cause of periodic cataclysmic Earth changes. It is the fact that our planet is tilted on its axis toward the sun. It is that coincidence that could be the cause of the repeat disasters.

The 23.5° list of the Earth on its axis means that at every solstice one pole is closer to the sun than the other. The accumulating ice on the poles means that year on year the differential in the gravitational pull from the sun on the poles during the solstice would grow stronger. Eventually, during a time of global warming, caused by energy rising from the core of the earth, the floating crust could quite literally roll 'head over heels'. This would also explain why the crust always slides over the poles when it slips. If the vortex model is correct and as we are currently in a period of global warming, we and our civilisation could be facing an imminent global catastrophe on an epic scale.

The attempt to account for global warming in terms of the industrial build up of carbon dioxide in the atmosphere is only a theory. If global warming is caused, rather, by the effect of a gravity geyser at the core centre of the Earth, then we could account for the predictions in science of an imminent magnetic pole reversal. No one knows how or why the magnetic poles reverse. Earth sciences demonstrate they have done so many times before. In his theory Hapgood spoke of the crust moving over the poles as an account for the periodic magnetic pole shifts but he did not talk about pole reversals.[6]

Hapgood suggested that the Earth's crust shifts just 30° at a time. The evidence of magnetic pole reversals suggests that the crust may roll more than 180°. The Earth's magnetism, generated by its massive iron core, may be weakened by the heat released by a gravity geyser but not reversed. However, the theory of the crust sliding over the poles provides a plausible explanation for the magnetic pole reversals. Once the crust begins to slide it may gain momentum and roll 'head over heels' coming to a standstill over opposite poles.

Were the crust of the Earth to roll over the inner layers it would generate heat by friction. Should the roll be to such an extent as to reverse the poles, the friction could generate sufficient heat to evaporate oceans and burn off practically everything on the land surface. The earthquakes caused by such a movement would be universal and off the Richter scale. Earthquakes, volcanic eruptions and an all consuming fire followed by a deluge, caused by ocean evaporation then followed by condensation when the Earth cools, would be major consequences of a *crust-roll* pole reversal.

If the crust were to reverse its position over the poles, any people left on the planet, after the cataclysm, would see the sun reverse its direction in the sky. After a pole reversal the sun would appear to rise in the West and set in the East.

According to ancient records the sun has reversed its direction in the sky many times before. They tell that the sun regularly alternates between rising in the East and setting in the West and rising in the West and setting in the East. In the Koran, Muslims are warned that a day is coming when humanity and civilisation will be mostly destroyed and the sun will again rise in the West.

6 www.scientificamerican.com/article/earth-s-magnetic-field-flip-could-happen-sooner-than-expected

CHAPTER 3
SUNRISE IN THE WEST

In *Worlds in Collision,*[1] Immanuel Velikovsky presents extraordinary accounts from ancient history that reveal a similar story of repeated global catastrophes in the folk memory of practically every culture on Earth. If the ancient records he unearthed are to be believed the world would appear to have been through repeated disasters and each time the sun appears to have reversed its direction in the sky.

Velikovsky discovered in Herodotus' *second book of history,* a reference to a conversation with Egyptian priests in which they asserted that since Egypt became a kingdom:

"...four times in this period the sun rose contrary to his wont; twice he rose where he now sets, and twice he set where he now rises."[2]

A Latin author of the first century, Pomponius Mela, commented:

"The Egyptians pride themselves on being the most ancient people in the world. In their authentic annals one may read that the course of the stars has changed direction four times and the sun has set twice in that part of the sky where it rises today."[3]

The *Papyrus Harris* records a cosmic upheaval when:

"...the south becomes north and the Earth turns over."[4]

The *Papyrus Ipuwer* stated:

1 Velikovsky I., *Worlds in Collision,* Victor Gollancz Ltd., 1950
2 Herodotus, *Bk ii 142* (transl. A.D. Godley 1921)
3 Mela Pomponius, *De Situ Orbis. i. 9. 8.*
4 Lange H., *Der Magische Papyrus Harris* Danske Videnskabernes Selskab, (p58) 1927

"The land turned round as does a potter's wheel and the Earth turned upside down."[5]

In the *Ermitage papyrus* reference is made to:

"...a catastrophe that turned the Earth upside down."[6]

The sun was known in ancient Egypt as *Harakhte.* There can be no doubt that the catastrophe was associated with a reversal in the direction of the sun because elsewhere in the Ermitage papyrus there is reference to the original direction of sunrise:

"Harakhte he riseth in the West."[7]

Texts found in the pyramids say the luminary ceased to live in the occident and shines anew in the orient.[8] Velikovsky went on to claim in the tomb of Senmut, architect of Queen Hatshepsut, there is a panel on the ceiling showing the celestial sphere with the signs of the zodiac and other constellations in a reversed orientation of the southern sky where north is exchanged for south and east for west.[9] The simplest conclusion to draw from this anomaly is that this was how the night sky originally appeared.

Velikovsky cited Plato in *Politicus*:

"I mean the change in the rising and setting of the sun and the other heavenly bodies, how in those times they used to set in the quarter where they now rise, and used to rise where they now set."[10]

5 Lange H., German translation of *Papyrus Ipuwer 2:8, (pp 601-610)* Sitzungsberichte Preuss. Akad. Der Wissenschaften 1903

6 Gardiner, *Journal of Egyptian Archeology I, (1914);* Cambridge Ancient History I, 346

7 Breasted, *Ancient Records of Egypt III, Sec 18.*

8 Speelers L., *Les Texts des Pyramides I, 1923*

9 Pogo A., *The Astronomical Ceiling Decoration in the Tomb of Senmu, Isis (p.306) 1930*

10 Plato, *Politicus (transl. H.N.Fowler, pp 49 and 53, 1925)*

The reversal of the sun in the sky was never a peaceful event, Plato continued in *Politicus*:

"There is at that time great destruction of animals in general and only a small part of the human race survived."

Velikovsky found in the drama *Thyestes* by Seneca a powerful description of what happened when the sun turned backward in the morning sky,[11] which Plato detailed in *Timaeus*:

"...a tempest of winds...alien fire...immense flood which foamed in and streamed out...the terrestrial globe engages is all motions, forwards and backwards and again to right and to left and upwards and downwards, wandering every way in all the six directions."[12]

In the sacred Hindu book, the *Bhagavata Purana,* the four ages or *Yugas* are described, each ending in a cataclysm in which mankind is almost destroyed by fire and flood, earthquake and storm.[13] These age changing events are documented by almost every culture. Hesiod, one of the earliest Greek authors, wrote about four ages and four generations of men destroyed by the wrath of the planetary gods.[14] He described the end of an age as:

"The life giving Earth crashed around in burning...all the land seething and the oceans...it seemed as if Earth and wide heaven above came together; for such a mighty crash would have arisen if Earth were being hurled to ruin, and heaven from on high were hurling her down."[15]

The Persian prophet Zarathustra spoke of:

11 Seneca, *Thyestes II,* (transl. F.J.Miller 794 ff)
12 Plato, *Timaeus,* (transl. Bury, 1929)
13 Moor E.,*The Hindu Pantheon* 1810
14 Hesiod, *Works and Days* (transl. H. Evelyn-White 1914)
15 Hesiod, Theogony (transl. H. Evelyn-White 1914)

"...signs, wonders and perplexity which are manifest on the Earth at the end of each age."[16]

The Chinese call the perish of ages 'Kis' and number, ten Kis from the beginning of their known world until Confucius.[17] In the ancient Chinese encyclopedia, 'Sing-li-ta-tsiuen-chou' the general convulsions of nature are discussed. Because of the periodicity of these convulsions the Chinese regard the span of time between two catastrophes as a 'great year'. As during a year, so during a world age, the cosmic mechanism winds itself up and:

"...in a general convulsion of nature, the sea is carried out of its bed, mountains spring out of the ground, rivers change their course, human beings and everything is ruined, and the ancient traces effaced."[18]

An ancient tradition of world ages ending in catastrophe is persistent in the Americas amongst the Incas,[19] the Aztecs and the Maya.[20] The Maya have a *Long Count* between global catastrophes similar to the *Chinese Great Year*.

In her book *Approaching Chaos* on global catastrophes in ancient times,[21] Lucy Wyatt suggested a date for Noah's Deluge as seventy nine years after the beginning of the *Long Count,* which ended in 2012. Her research indicated that cataclysms occur close to the target date of a Long Count, or a Great Year rather than on it. That would fit with the principle in science of *margin of errors*. In a cycle of global cataclysms that are millennia apart, one would expect a cataclysmic event to occur sometime within a margin of a few decades of a predicted date.

16 Muller F., *Pahlavi Texts: The Sacred Books of the East,* 1880
17 Murray ed. *Historical and Descriptive Account of China,* 1836
18 Schlegel G., *Uranographie Chinoise,* Wou-foung, 1875
19 Alexander H., *Latin American Mythology,* 1920
20 Humbolt A von, *Researches II,15.*
21 Wyatt L., *Approaching Chaos,* O Books 2009

The Maya described each age ending with earthquakes at the solstice. The Maya reference to a worldwide catastrophe of global earthquakes at the solstice supports the Hapgood model of crust slips being related to the accumulation of ice on the poles. According to my model the difference in gravitational pull from the sun on the ice on the poles, due to the 23.5° list of the Earth on its axis, would cause a floating crust, during a period of global warming, to be most prone to rolling over at the solstice. The greatest likelihood would be December 21st as that would be when the ice on the Antarctic would experience the strongest gravitational pull from the sun. Hapgood pointed out that ice accumulating on a land mass was more prone to cause a crust shift than ice floating on the sea. The Maya prediction fits the science.

Historical records from every continent report that the world has fallen over the poles at least four times in the memory of mankind. A major part of stone inscriptions found in the Yucatan refer to this type of world catastrophe. The most ancient of these *katun* calendar stones of Yucatan refer to great catastrophes, at repeated intervals, convulsing the American continent. The indigenous nations of the Americas have a preserved memory of these ancient historical events.[22] In the chronicles of the Mexican kingdom it was written:

"The ancients knew that before the present sky and earth were formed, man was already created and life had manifested itself four times."[23]

The sacred Hindu books, the *Ezour Vedam* and the *Bhaga Vedam* share the scheme of expired ages known as Yugas, the fourth being the present. They differ only in the time ascribed to

22 Brasseur de Bourbourg C., *S'il existe des Sources de l'histoire primitive du Mexique dans les monuments égyptiens*, 1864

23 Brasseur de Bourbourg C., *Histoire des nations civilisées du Mexique I*, 53, 1857-1859

each age.[24] The Buddhist text *Visuddhi-Magga* also describes seven ages, each terminated by world catastrophes.[25]

A tradition of successive creations and catastrophes is found in Hawaii.[26] On the islands of Polynesia there were nine ages recorded and in each age a different sky was above the Earth.[27] Icelanders believed that nine worlds went down in a succession of ages, a tradition contained in the Mavarian text *Edda.* [28]

There are seven ages referenced in the rabbinical tradition of *Creation*:

"Already before the birth of our earth, worlds had been shaped and brought into existence, only to be destroyed in time. This Earth too was not created in the beginning to satisfy the divine plan. It underwent reshaping, six consecutive remouldings. New conditions were created after each of the catastrophes...we belong to the seventh age.[29]

According to the rabbinical authority, Rashi, who lived in North London, Hebrew tradition knew of periodic collapse of the firmament, one of which occurred in the days of the Deluge.[30]

The Jewish philosopher Philo wrote:

"Great catastrophes changed the face of the Earth. Some perished by deluge, others by conflagration.[31]

In Isaiah 24:1 is written:

24 Volney C., *New Researches on Ancient History* (p.56) 1856
25 Warren H., *Buddhism in Translations* 1896
26 Dixon R., *Oceanic Mythology*, 1916
27 Williamson R., *Religious and Cosmic Beliefs of Polynesia* 1933
28 Völuspa, *The Poetic Edda,* (transl. H. Bellows 1923)
29 Ginzberg L., *Legends of the Jews*, 1925
30 Rashi, *Commentry to Genesis* 11:1
31 Philo, *Moses II, x, 53*

"Behold the Lord maketh the earth empty, and maketh it waste, and turneth it upside down, and scattereth abroad the inhabitants thereof."

Reports of the Earth turning upside down over the poles are difficult to take seriously but coming from so many and diverse sources they are also difficult to dismiss. Graham Hancock provides evidence of repeated Earth changes and many lost civilisations in his compelling books, especially in *Fingerprints of the Gods.*[32]

If, as described by Charles Hapgood in the *Earth's Shifting Crust,*[33] the crust of the Earth were to shift over the outer mantle then that could cause the vivid events described by the ancients without the entire globe going for a tumble as imagined by Velikovsky. In the Hapgood hypothesis the poles don't change. It is the crust turning over the poles that causes the apparent reversal which would cause the change in direction of the sun in the sky. This is because if the pole positions on the crust reversed, by its rolling over the actual poles of the turning globe, the east-west orientation on the crust would also reverse. A displacement of 180° or more, of the Earths crust over the poles of the planet, would account for the claims in the ancient records that the sun periodically reverses its direction in the sky. Maybe the South Pole was situated under Hudson Bay ten thousand years ago. Maybe when the rivers were flowing on the continent, now known as Antarctica, it was in the Northern hemisphere.

A reverse displacement of the Earth's crust over the poles would cause shock waves that would lead to unprecedented earthquakes and tsunamis throughout the world. There would also be extreme heat caused by friction as the crust rolls over. This would cause conflagrations and violent storms described by the ancients. Unprecedented levels of evaporation of the

32 Hancock G. *Fingerprints of the Gods,* Century Books, 2001

33 Hapgood, C., *Earth's Shifting Crust: A Key to Some Basic Problems of Earth Science,* Pantheon Books, 1958

oceans followed by condensation when the Earth settles and cools down would account for the deluges recorded in so many ancient texts. All of this could account for the worldwide catastrophes reported by the ancients, of earthquake and storm, fire and flood which accompany the reversal of direction of the sun in the sky.

At the beginning of the 21st Century we are in the midst of an accelerating phase of global warming, so it is highly likely we are approaching another crust shift and consequent pole reversal. In 2014, the European Space agency predicted an imminent magnetic pole reversal but by imminent they meant any time in a few thousand years. However, in *Scientific American* an article suggested a reversal of the poles could happen much sooner than they expected.[34]

The hypothesis proposed by Charles Hapgood and endorsed by Albert Einstein in *Earth's Shifting Crust* links a number of seemingly disconnected events that return to destroy human civilisations and reduce populations. My *gravity geyser* model accounting for the periodic rise of heat in the Earth, provides an account for the energy that could cause periodic global catastrophes accompanied by reversals of the direction of sunrises and sunsets recorded in the historical annals. I call the periodic releases of gravity energy from the core of the planet, the *heart beat of the Earth*. Annihilation energy rising periodically from the centre of the Earth could explain the Hindu tradition of Shiva the destroyer awakening to end a *Yuga* and herald the dawn of a new age. So many great civilisations have vanished without trace. Maybe that is because none can survive a beat of the planetary heart.

34 www.scientificamerican.com/article/earth-s-magnetic-field-flip-could-happen-sooner-than-expected

CHAPTER 4
THE ACADEMY OF ANGELS

In the paradigm of super-physics, where science could lead the way for human awakening, we can begin to understand that each human incarnation offers the Universe an opportunity to become aware of itself. We receive the gift of life. In return we give back to the Universe our experiences and our service.

Just as we go through phases of education to prepare us for life on Earth, kindergarten, primary school, secondary school and university, so it would seem we go through four similar phases of education – the *four ages of mankind* – to prepare us for life and service in the higher worlds as divine beings.

According to the ancient Vedic scriptures of India, which form the basis of the Hindu religion, there are four Yugas or great ages of training for humanity. Beginning with the Sat Yuga, a golden age of light, unity and truth, at the start of each cycle and culminating with the Kali Yuga, a most brutal age of darkness and discord, at the end of the cycle.

This notion of four archetypal ages of growth and learning was also understood in ancient Greece and Rome. To quote Thomas Bulfinch:[1]

"The first age was an age of innocence and happiness called the Golden Age. Truth and right prevailed though not enforced by law, nor was there any magistrate to threaten or punish. The forest had not yet been robbed of its trees to furnish timbers for vessels nor had men built fortifications round their towns. There were no such things as swords, spears or helmets. The Earth brought forth all things necessary for man, without his labour in ploughing or sowing...

1 Bulfinch T. *The Golden Age of Myth & Legend,* Wordsworth Reference, 1993

"Then came the Silver Age, inferior than the Golden but better than the Brass... For the first time men had to endure the extremes of heat and cold, and houses became necessary. Caves were the first dwellings, and leafy coverts of the woods, and huts woven of twigs. Crops would no longer grow without planting. The farmer was obliged to sow the seed with a toiling ox to draw the plough.

"Next came the Brass Age, more savage of temper and readier to the strife of arms, yet not altogether wicked. The hardest and worst was the Iron Age. Crime burst in like a flood; modesty, truth and honour fled. In their places came fraud and cunning, violence, and the wicked love of gain. Then seamen spread sails to the wind, and trees were torn from the mountains to serve for keels of ships, and vex the face of the ocean. The Earth, which until now had been cultivated in common, began to be divided off into possessions. Men were not satisfied with what the surface produced, but must dig into its bowels, and draw forth from thence the ores of metals. Mischievous iron and more mischievous gold were produced. War sprang up, using both as weapons; the guest was not safe in his friend's house; and sons-in-law and fathers-in-law, brothers and sisters, husbands and wives could not trust one another. Sons wished their fathers dead, that they might come to the inheritance; family love lay prostrate. The Earth was wet with slaughter..."

The idea of a 'fall from grace' has been transmitted through myths and stories across many of the world's cultures. An expulsion from a heavenly paradise and an adventure of descent from innocence and cooperation into a world of despairing separation from our divine heritage is the mythological cornerstone of our western civilisation. These stories, rather than being folk myths, may well contain an encoded message about the growth cycle of humanity.

It could be that the four Yugas, or the four ages as described in great detail in the Vedic scriptures, are the four growth periods for souls incarnating on Earth, which acts as an *academy for angels*. From an age of gold, or golden age, through to the age of Iron, it could be that a soul moves through educational phases from kindergarten to university. It is possible that everything that happens on Earth is part of a curriculum saga designed to build strength of spirit particularly through difficulty, challenge and adversity. In one of my songs I wrote:

We are the gods here in human form,
Here to overcome through battle, pain and storm,
That tenacious hold onto power, greed and pride,
That prevents us, as gods, from evolving on high.

Some of us may be angels already. We may have voluntarily descended from a heavenly state down to the gravity plane of Earth to serve as teachers and learn more lessons. Some angels may have fallen to Earth to be reformed. In this sense, the *Earth Academy of Angels* may be acting as a crucible of fire for angels where dross can be separated from gold and spiritual coal can be transformed into diamond through the heat and pressure of adversity.

The ancients believed the catastrophic pole shifts were vengeance wreaked upon sinful humanity. However, the recurrent disasters that afflict the Earth are inevitable terrestrial changes that are not linked in any way to human behaviour. They are the consequence of the Earth having been knocked off its axis by 25.5° at some point in its history and a black hole *gravity geyser* located at its core.

Neither is the Kali Yuga or Age of Iron bad for us. The Kali Yuga represents the age when evil and ignorance are rampant on the Earth but we need to experience evil as well as good. The greed and selfishness, materialism and carnage in the last age of the Yuga sequence is a vital part of our angelic education.

There is no judgment of humanity, individual or en masse. None of us are in a position to judge. Now is the time for us to let go of judgment between good and evil. For example, many people are anxious to learn about a group of supposedly deviant celestial beings called *Archons*. The Archons are referred to in Gnostic teachings in parchments discovered in Nag Hammadi, Egypt, in 1947. The teachings date back hundreds of years B.C.E.

The Gnostics tell us that Archons maintain the Earth as a prison for human souls. Archon is a Greek word that literally means *prince* or *ruler*. In fascinating detail the Gnostic treatises tell us that the Archons are servants of a *Demiurge*, an intermediary celestial entity that stands between the human race and a transcendent God. Only through Gnosis – direct revelation of the Knowledge of the soul's divine nature – can a soul be liberated from enmeshment in the material dimension and return to the upper worlds of the Divine.

It is my suggestion that Archons play an important role and have a purpose so integral to the maintenance and order of the Universe that without them the *Earth Academy for Angels* would be in disarray and could attract cosmic disrepute. Archons could be regarded as the equivalent of teachers in a boarding school. The all important role they fulfill is to challenge us to strengthen our spirits through difficulty, challenge and adversity.

As adversarial forces, the Archons train us through antagonism and suffering. They foster our spiritual constitution with a view to strengthening our capacity for illumination and decision making at the time of our final examination at the moment of decision when we are offered the opportunity for ascension. They set up temptations on the physical plane to addictions and attachments. This furthers the end and to examine our progress through continuous assessment. This might explain why, in the *Book of Enoch*, the Archons – fallen angels – are called *The Watchers*. [2]

The job of teachers in a boarding school is also to stop pupils absconding. An additional role of these primordial planetary spirits is to prevent juvenile angels from escaping the *Earth Boot Camp for Angels*. As with schoolmasters they use whatever measures are necessary to maintain and keep pupils within the boundaries of the academy until graduation, when parents come to collect them. Teachers are sometimes vilified by pupils as monsters, cold-hearted devils, evil captors and suchlike. By the same token, Gnostics regarded Archons as 'rulers of darkness'.

The head of the Archons; the equivalent of headmaster in the boarding school analogy – sometimes known as *Satan* or the Devil – is also known as the gatekeeper of *the beyond the beyond*.[3]

2 Enoch, *The Book of Enoch*, Defender, 2017
3 Ash D. *The Role of Evil in Human Evolution*, Kima Global Publishing, 2007

The beyond is the *hyperphysical* 'fourth dimension' and the 'beyond the beyond' is the *superphysical* 'fifth dimension'. His role, representing the Archons, is not only to tempt and test us to bring out to the fore our inner hidden tendencies, but also to ensure we don't break out into the Universe at large until we have passed the most stringent test of quarantine; the ability to be unconditional in our love and in our judgments. Humans have the propensity to be planetary pathogens. We may despise Satan but the Universe relies on him to maintain galactic health and safety.

Satan is an anagram of *Sanat*. Sanat Kumara (Kumara means 'prince') is the first born son of Brahma. Sanat Kumara is also known as the *Ancient of Days* which identifies him as YHVH: *Yodheva* the God of the Old Testament. The head of the Archons is not evil. He is just doing his job.

A soul that aspires to grow, evolve and develop the treasury of light within him or herself will come up against subjugating obstacles and forces. Referred to by Paul the Apostle as the authorities of the darkness – *skotos: obscurity* – we are told by Paul that the supreme contest of a soul is not against flesh and blood but rather against the authorities of the Universe and the spirits of obscurity, acting somewhat like antagonistic muscles:

"For we wrestle not against flesh and blood, but against principalities, against powers, against the rulers of the darkness of this world, against spiritual wickedness in high places."

Paul from Ephesians 6:12

The Archons are simultaneously intelligences in the outer cosmos, active without and in the inner cosmos. Yet, as with all serious spiritual work, the primary concern is developing inner strength and wisdom to overcome obstacles and limitations on the inner plane. If we cannot conquer the forces of spiritual inertia, and apathy within ourselves, then we'll never conquer them without.

An esoteric interpretation of the ruling Archonic forces is to understand them as portions of the *Cosmic Being* trapped within the cage of the impure and ignorant mind. These internal forces have been personified and mythologised in world traditions as *Pluto, Hades, Ahriman* and *Lucifer* with his fallen angels.

At the point of ascension, when the super physical 'parent gods or angels' that seeded us here in the first place, return to the third and fourth dimensions of the Earth to collect us – we are their offspring – the Archons will not stand in their way. They will greet us on the other side of the doorway of light. They will guide us through the flame of ascension. They will witness our graduation from the Earth as an angel. They will join us in the celestial celebrations. The Archons will not hold us to the Earth planes if we choose to enter the fifth dimension via the doorway of light. However, up to that point they will test us. For example they have agents warning us to avoid the light, that it will trap us. They do this to test our fears because our fears have the power to block the path to our release from the terrestrial plane.

Everything happens for a purpose. There are no mistakes. When, according to Gnostic thought, *Sophia* broke away from the spiritual planes to form the physical realms of the Earth, when as the wisdom embodiment of Gaia she became the *anima mundi*, the soul of the Earth, it was not in error. She was creating a womb. She was establishing the *Earth Womb of Angels*. Sophia was also building an angelic school; the *Earth Academy for Angels*. She did not cause our fall; she was founding our academy. Sophia is our divine mother. She is also the equivalent of school governor in the boarding school analogy. Educating us as up and coming angels through endless trials and tribulations, she is ultimately responsible for forging us as gods.

Sophia can be understood as an aspect of *Adi Para Shakti* (the Supreme Being in Hindu mythology who gave birth to the trinity of Brahma-Vishnu-Shiva). In the Yogic tradition, *Adi Para Shakti* is the origin of life and all that is. I believe an aspect of *Adi Para Shakti* broke away from plasmic perfection as Sophia, the embodiment of wisdom, to set up our world. Sophia oversees our birth as baby angels, birthing through the chemistry of atomic matter, to learn fundamental lessons about life and the Universe. This cosmic birthing process is enabled through the amnesia of our divine nature and through our perceived limitations and separation.

In my philosophy, if Sophia brought Archons in her wake, She brought them along as boot camp guardians, and tough, uncompromising adversarial teachers. There is no evil in this. It is for our ultimate good. There is nothing for us to fear.

It is easy to blame others, like the Archons, for our plight and our pain, and we are quick to judge between good and evil yet evil as well as good has its place in the evolution of humankind.[4] Everyone has a purpose and everything has a place in the Divine plan to teach us and put us through our paces. Fallen angels that use their powers to dominate others serve in educating us as angels about the darker side of nature. Incarnating as, or possessing people as psychopaths, despots or tyrants, or even as members of our family or people we are in relationship with, they increase our suffering for our ultimate greater good. There is no absolute evil in them. They work for our ultimate good by acting as mirrors for us. If we see evil it is only a reflection of ourselves. If we see good that too is our own reflection.

Coal carbon can form diamonds but only through the fearsome forces of heat and intense pressure. Gems are polished by tumbling with grit, gold is purified by fire. So it is that we human souls are forged through anguish and torment, by forces perceived as evil, for the express purpose that we come into a perfection of our own making.

We may complain but as Rumi said, "If you are irritated by every rub, how will you be polished?"

The cycle of ages from gold to iron, from goodness to evil and back again, finds symbolic representation in the Yin Yang symbol. The Chinese principle of Yin and Yang is that the greatest good carries the seed of evil and the greatest evil carries the seed of good, then when each polarity comes to its extreme it flips to the opposite.

Most of us were seeded into human incarnation at the beginning of a golden age around twenty three thousand years ago. The golden age was sweet and harmonious for us as angel newborns. We found ourselves in the equivalent of pre-school or

4 Ash D. *The Role of Evil in Human Evolution*, Kima Global Publishing, 2007

kindergarten where ascended humans from previous cycles had returned to the planet as 'the shining ones, the gods or the divines', to conceive, birth and guide us into the terrestrial system of angel education. From there we moved on to a silver age, the equivalent of primary school, then onto a brass age, the equivalent of secondary or high school until we entered the iron age, or Kali Yuga, the equivalent of university where we have been learning to overcome every imaginable humiliation, fear and terror until we know only love. If we have learnt the lesson that love prevails, then through ascension, we can graduate into the Universe at large, and return to Earth, after its purification, in the new age, to seed the next generation of angels; our own angelic offspring.

Fortunate are those who pass the tests at the end of the course and let go of everything to enter the doorway of light and graduate in ascension. We have come such a long way to reach this threshold. Let us not throw it all away for the sake of money, attachments, possessions, fears or petty issues. Everything has to go someday. Warnings have resounded down the ages and are all around us now, of what is to come and of the ascension opportunity for us all. One of the great heralds of this time was Peter Deunov of Bulgaria.

CHAPTER 5
THE PETER DEUNOV PROPHECY

Albert Einstein said, "The whole world bows down before me; I bow down before the master Peter Deunov from Bulgaria".

Pope John XXIII described Peter Deunov as, "The greatest philosopher on Earth."

Days before he died, Peter Deunov (1864-1944), came out with this astounding prophecy:

Peter Deunov

"During the passage of time, the consciousness of man traversed a very long period of obscurity. This phase, which the Hindus call 'Kali Yuga', is on the verge of ending. We find ourselves today at the frontier between two epochs: that of Kali Yuga and that of the New Era that we are entering.

"A gradual improvement is already occurring in the thoughts, sentiments and acts of humans, but everybody will soon be subjugated to divine Fire, that will purify and prepare them in regards to the New Era. Thus man will raise himself to a superior degree of consciousness, indispensable to his entrance to the New Life. That is what one understands by 'ascension'.

"Some decades will pass before this Fire will come, that will transform the world by bringing it a new moral. This immense wave comes from cosmic space and will inundate the entire earth. All those that attempt to oppose it will be carried off and transferred elsewhere.

"Although the inhabitants of this planet do not all find themselves at the same degree of evolution, the new wave will be felt by each one of us. And this transformation will not only touch the Earth, but the ensemble of the entire Cosmos.

"The best and only thing that man can do now is to turn towards God and improve himself consciously, to elevate his vibratory level, so as to find himself in harmony with the powerful wave that will soon submerge him.

"The Fire of which I speak, that accompanies the new conditions offered to our planet, will rejuvenate, purify, reconstruct everything: the matter will be refined, your hearts will be liberated from anguish, troubles, incertitude, and they will become luminous; everything will be improved, elevated; the thoughts, sentiments and negative acts will be consumed and destroyed.

"Your present life is a slavery, a heavy prison. Understand your situation and liberate yourself from it. I tell you this: exit from your prison! It is really sorry to see so much misleading, so much suffering, so much incapacity to understand where one's true happiness lies.

"Everything that is around you will soon collapse and disappear. Nothing will be left of this civilization nor its perversity; the entire earth will be shaken and no trace will be left of this erroneous culture that maintains men under the yoke of ignorance. Earthquakes are not only mechanical phenomena, their goal is also to awaken the intellect and the heart of humans, so that they liberate themselves from their errors and their follies and that they understand that they are not the only ones in the Universe.

"Our solar system is now traversing a region of the Cosmos where a constellation that was destroyed left its mark, its dust. This crossing of a contaminated space is a source of poisoning, not only for the inhabitants of the earth, but for all the inhabitants of the other planets of our galaxy. Only the suns are not affected by the influence of this hostile environment. This region is called "the thirteenth zone"; one also calls it "the zone of contradictions". Our planet was enclosed in this region for thousands of years, but finally we are approaching the exit of this space of darkness and we are on the point of attaining a more spiritual region, where more evolved beings live.

"The earth is now following an ascending movement and everyone should force themselves to harmonize with the currents of the ascension. Those who refuse to subjugate themselves to this orientation will lose the advantage of good conditions that are offered in the future to elevate themselves. They will remain behind in evolution and must wait tens of millions of years for the coming of a new ascending wave.

"The Earth, the Solar system, the Universe, all are being put in a new direction under the impulsion of Love. Most of you still consider Love as a derisory force, but in reality, it is the greatest of all forces! Money and power continue to be venerated as if the course of your life depended upon it. In the future, all will be subjugated to Love and all will serve it. But it is through suffering and difficulties that the consciousness of man will be awakened.

"The terrible predictions of the prophet Daniel written in the bible relate to the epoch that is opening. There will be floods, hurricanes, gigantic fires and earthquakes that will sweep away everything. Blood will flow in abundance. There will be revolutions; terrible explosions will resound in numerous regions of the earth. There where there is earth, water will come, and there where there is water, earth will come. God is Love; yet we are dealing here with a chastisement, a reply by Nature against the crimes perpetrated by man since the night of time against his Mother; the Earth.

"After these sufferings, those that will be saved will know the Golden Age, harmony and unlimited beauty. Thus keep your peace and your faith when the time comes for suffering because it is written that not a hair will fall from the head of the just. Don't be discouraged simply follow your work of personal perfection.

"You have no idea of the grandiose future that awaits you. A New Earth will soon see day. In a few decades the work will be less exacting, and each one will have the time to consecrate spiritual, intellectual and artistic activities. The question of rapport between man and woman will be finally resolved in harmony; each one having the possibility of following their aspirations. The relations of couples will be founded on reciprocal respect and esteem. Humans will voyage through the different planes of space and break through intergalactic space. They will study their functioning and will rapidly be able to know the Divine World, to fusion with the Head of the Universe.

"The New Era is that of the sixth race. Your predestination is to prepare yourself for it, to welcome it and to live it. The sixth race will build itself around the idea of Fraternity. There will be no more conflicts of personal interests; the single aspiration of each one will be to conform himself to the Law of Love. The sixth race will be that of Love. A new continent will be formed for it. It will emerge from the Pacific, so that the Most High can finally establish His place on this planet.

"The founders of this new civilization, I call them 'Brothers of Humanity' or also 'Children of Love'. They will be unshakeable for the good and they will represent a new type of men. Men will form a family, as a large body, and each people will represent an organ in this body. In the new race, Love will manifest in such a perfect manner, that today's man can only have a very vague idea.

"The earth will remain a terrain favourable to struggle, but the forces of darkness will retreat and the earth will be liberated from them. Humans seeing that there is no other path will engage themselves to the path of the New Life, that of salvation. In their senseless pride, some will to the end hope to continue on earth a life that the Divine Order condemns, but each one will finish by understanding that the direction of the world doesn't belong to them.

"A new culture will see the light of day, it will rest on three principal foundations: the elevation of woman, the elevation of the meek and humble, and the protection of the rights of man.

"The light, the good, and justice will triumph; it is just a question of time. The religions should be purified. Each contains a particle of the Teaching of the Masters of Light, but obscured by the incessant supply of human deviation. All the believers will have to unite and to put themselves in agreement with one principal, that of placing Love as the base of all belief, whatever it may be. Love and Fraternity that is the common base! The earth will soon be swept by extraordinary rapid waves of Cosmic Electricity. A few decades from now beings who are bad and lead others astray will not be able to support their intensity. They will thus be absorbed by Cosmic Fire that will consume the bad that they possess. Then they will repent because it is written that "each flesh shall glorify God".

"Our mother, the earth, will get rid of men that don't accept the New Life. She will reject them like damaged fruit. They will soon not be able to reincarnate on this planet; criminals included. Only those that possess Love in them will remain.

"There is not any place on earth that is not dirtied with human or animal blood; she must therefore submit to a purification. And it is for this that certain continents will be immersed while others will surface. Men do not suspect to what dangers they are menaced by. They continue to pursue futile objectives and to seek pleasure. On the contrary those of the sixth race will be conscious of the dignity of their role and respectful of each one's liberty. They will nourish themselves exclusively from products of the vegetal realm. Their ideas will have the power to circulate freely as the air and light of our days.

"The words "If you are not born again" apply to the sixth race. Read Chapter 60 of Isaiah it relates to the coming of the sixth race, the Race of Love.

"After the Tribulations, men will cease to sin and will find again the path of virtue. The climate of our planet will be moderated everywhere and brutal variations will no longer exist. The air will once again become pure, the same for water. The parasites will disappear. Men will remember their previous incarnations and they will feel the pleasure of noticing that they are finally liberated from their previous condition.

In the same manner that one gets rid of the parasites and dead leaves on the vine, so act the evolved Beings to prepare men to serve the God of Love. They give to them good conditions to grow and to develop themselves, and to those that want to listen to them, they say: "Do not be afraid! Still a little more time and everything will be all right; you are on the good path. May he that wants to enter in the New Culture study, consciously work and prepare."

Thanks to the idea of Fraternity, the earth will become a blessed place, and that will not wait. But before, great sufferings will be sent to awaken the consciousness. Sins accumulated for thousands of years must be redeemed. The ardent wave emanating from On High will contribute in liquidating the karma of peoples. The liberation can no longer be postponed. Humanity must prepare itself for great trials that are inescapable and are coming to bring an end to egoism.

Under the earth, something extraordinary is preparing itself. A revolution that is grandiose and completely inconceivable will manifest itself soon in nature. God has decided to redress the earth, and He will do it! It is the end of an epoch; a new order will substitute the old, an order in which Love will reign on earth.

EPILOGUE

If you are inspired by *Awaken* please circulate copies to your family and friends. We are living in times of prophecy. The liberating process of ascension into higher worlds is for those who elect themselves for awakening. It is not an automatic process; it depends on choice. The ascension opportunity is available for all human beings dwelling on the Earth but we have to seize the opportunity. We have to choose spirituality over materialism. We will all be offered the same opportunity to ascend by the appearance of a doorway of light. In order to ascend we need to be able to step into the light fearlessly and leave everything in this world behind without a moment's hesitation. We can start preparing for that eventuality by being more spiritually aware, by caring and sharing, by spreading the message of ascension, by living out of love for others, by practicing compassion for all that lives, by caring for the Earth and by acting always in support of the upliftment of humanity.

If we can accept the gift of despair and devastation we will receive the gift of divinity. God is the Gift of Divinity we receive on our ascension.

Shivali

However men approach me, even so do I welcome them, for the paths men take from every side are mine.

The Bhagavad Gita

A VEDIC PRAYER FOR LIBERATION AND ASCENSION

THE PAVAMANA ABHYAROHA MANTRA

Pavamana means 'being purified'
Abhyaroha means 'ascending'

Om asato ma sad gamaya,
Om, from falsehood lead me to truth,

tamaso ma jyotir gamaya,
From darkness lead me to the light,

m'tyor ma amta gamaya,
From death lead me to immortality,

Om shanti~ shanti~ shanti hi~
Om peace peace peace.

BIBLIOGRAPHY

Allan D. *When the Earth Nearly Died* Gateway Books, 1995

Armstrong K. *Islam: A Short History,* Phoenix, 2001

Ash D. & Anna A. *The Tower of Truth,* CAMSPRESS, 1977

Ash D. & Hewitt P. *The Vortex: Key to Future Science,* Gateway Books, 1990

Ash D. *The New Science of the Spirit,* The College of Psychic Studies, 1995

Ash D. *Activation for Ascension,* Kima Global Publishing 1995

Ash D. *The Role of Evil in Human Evolution,* Kima Global Publishing 1995

Ash D. *The New Physics of Consciousness,* Kima Global, 2007

Ash D. *The Vortex Theory,* Kima Global Publishing, 2015

Ash D. *Continuous Living,* Kima Global Publishing 2015

Ash M. *Health, Radiation and Healing,* Darton, Longman & Todd, 1963.

Berkson W. *Fields of Force: World Views from Faraday to Einstein,* Rutledge & Kegan Paul, 1974

Bielek A. & Steiger B. *The Philadelphia Experiment,* Inner Light

Bloom W. *Working with Angels, Fairies and Nature Spirits,* Piatkus, 1998

Bulfinch T. *The Golden Age of Myth & Legend,* Wordsworth Reference, 1993

Burr H.S. *Blueprint for Immortality,* Neville Spearman, 1972

Calder N. *Key to the Universe: A Report on the New Physics,* BBC Publications, 1977

Capra F. *The Tao of Physics,* Wildwood House, 1975

Capra F. *The Turning Point,* Fontana, 1983

Cook N. *The Hunt for Zero Point,* Arrow books, 2002

Davidson J. *The Secret of the Creative Vacuum,* C.W.Daniel 1989

Dawkins R. *The Blind Watchmaker,* W. W. Norton & Co, 1986

Dawkins R. *The God Delusion,* Bantam Press, 2006

Edwards H. *Harry Edwards: Thirty Years a Spiritual Healer,* Jenkins 1968

Eldredge N. *The Triumph of Evolution and the Failure of Creationism,* Freeman & Co., 2000

Fritzsh H. *Quarks: The Stuff of Matter,* Allen Lane, 1983.

Gamow G. *Thirty Years that Shook Physics: The Story of Quantum Theory,* Dover Poblications 2003

Gardner L. *The Origin of God,* Dash House, 2010

Geng J., *Selecting the Right Acupoints: Handbook on Acupuncture Therapy* New World Press,China 1995

Good T. *Above Top Secret,* Sidgwick & Jackson, 1987

Gupta H. and Hahnemann S. *Medicine for the Wise: Hahnemann's Philosophy of Diseases, Medicines and Cures,* Create Space Independent Publishing Platform, 2014

Hapgood, C., *Earth's Shifting Crust: A Key to Some Basic Problems of Earth Science,* Pantheon Books, 1958

Haraldsson E. *Modern Miracles: The Story of Sathya Sai Baba: A Modern Day Prophet,* White Crow Books, 2013

Harari Y. N. *Sapiens: A Brief History of Humankind,* Vintage, 2014

Harari Y.N. *21 Lessons for the 21st. Century* Jonathan Cape 2018

Hawken P. *Blessed Unrest,* Viking, 2007

Hawking S. *A Brief History of Time,* Bantam, 1988

Hawking S. *Black Holes and Baby Universes,* Bantam, 1993

Heley Mark, *The Everything Guide,* Adams Media. 2009

Hoyle F. *The Intelligent Universe,* Michael Joseph, 1983

Huxley A. *The Doors of Perception,* Penguin Books, 1959

Jeans J. *The Mysterious Universe,* Cambridge Univ. Press, 1930

Keyes K. *The Hundredth Monkey,* Camarillo De Vorss & Co. 1984

Klein E. *The Crystal Stair,* Oughten House, 1992

Kuhn T. *The Structure of Scientific Revolutions,* University of Chicago Press, 1962

Leggett A.J. *The Problems of Physics,* Oxford Univ. Press, 1987

Lenz F. *Lifetimes,* Bobbs-Merrill, 1979

Levengood W.C. *Anatomical anomalies in crop formation plants,* Physiol. Plant 92, 356, 1994

McKenzie A. E. *A Second MKS Course in Electricity,* Cambridge University Press, 1968

McTaggart L. *The Field,* Harper & Collins, 2001

Moody R. *Life after Life,* Bantam Books, 1967

Nichols B. P. *The Montauk Project,* Sky Books, 1992

Ormond R. *Into the Strange Unknown* Esoteric Publishing 1959

Pagels H. *The Quantum Code,* Michael Joseph, 1982

Parnia S. *Erasing Death,* Harper One, 2013

Penwyche G. *The World of Angels* Bounty Books 2009

Popper K. *The Logic of Scientific Discovery,* Hutchinson, 1968

Ramacharaka Yogi, *An Advanced Course in Yogi Philosophy,* Fowler 1904, (Cosimo Classics Philosophy, 2005)

Redfield J. *The Celestine Prophesy* Transworld Publishers 1994

Rees M. *Just Six Numbers,* Weidenfield & Nicolson, 1999

Richards, et al, *Modern University Physics* Addison-Wesley 1973.

Rovelli C. *Reality is not what it seems to be,* Penguin, 2017

Sams G. *Sun of gOd,* Red Wheel Weiser, 2009

Sánchez-Ventura y Pascual F., *The Apparitions of Garabandal,* San Miguel, 1966

Sartori P. *The Wisdom of Near-Death Experience,* Watkins, 2014

Sartori P. *The Near Death Experiences of Hospitalized Intensive Care Patients: A Five Year Clinical Study,* Edwin Meller 2008

Schwartz J. *Sudden Origins – Fossils, Genes and the Emergence of Species*, John Wiley & Sons , 1999

Segre E. *Nuclei & Particles*, Benjamin Inc, 1964

Sheldrake R. *A New Science of Life*, Paladin Books, 1987

Sheldrake, Rupert *The presence of the past: Morphic resonance and the habits of nature*, Icon Books, 2011

Stevenson I. *Cases of the Reincarnation Type and Twenty Cases of Suggestive Reincarnation*, University of Virginia Press

Stokes D. F. *Voices In My Ear*, Futura Publications, 1980

Storm H. *My Descent into Death*, Clairview, 2000

Sundeen M. *The Man Who Quit Money*, Riverhead, 2012

Swedenborg E. *Arcana Coelestia*, Forgotten Books, 2008

Swedenborg E. *Heaven and Hell*, Swedenborg Foundation Publishers, 2010

Swedenborg E., *A Swedenborg Sampler:* Swedenborg Foundation Publishers, 2011

The Townsend Brown Foundation *The Biography of Thomas Townsend Brown & Project Winterhaven*

Velikovsky I. *Worlds in Collision*, Victor Gollancz, 1950

Von Buttlar J. *The Philadelphia Experiment*

Von Buttlar J.*The UFO Phenomenon*, Sidgwick & Jackson, 1979

Wilhelm R. *I Ching*, Routledge & Kegan Paul, 1951

Wyatt L. *Approaching Chaos*, O Books, 2009

INDEX

H

M

S

ABOUT THE AUTHOR

David Ash was born in Kent, England, in 1948. His father, Dr Michael Ash, a medical doctor researching the link between cancer and radioactivity, introduced David as a boy to nuclear physics, healing and nutrition. David started to develop the vortex theory while he was at Queens University of Belfast in 1968 and continued at Queen Elizabeth College, London University 1969 to 1972 and then at the College of St Mark and St John in Plymouth in 1973/74. On January 15th 1975 he presented his vortex theory from the Faraday Rostrum at the Royal Institution in the same auditorium where the leading physicist of the nineteenth century, Lord Kelvin, also presented the historic vortex theory for the atom. David then continued to develop the theory until it was published (with Peter Hewitt) as The Vortex: Key to Future Science by Gateway books in 1990; after which he set out on an international lecture tour between 1991-94 to promote the vortex theory and the associated prediction of ascension. In 1995 the College of Psychic Studies published The New Science of the Spirit, which included a prediction of the accelerating expansion of the Universe, subsequently confirmed experimentally by Saul Perlmutter in 1997. The vortex theory was revised and republished as The New Physics of Consciousness by Kima Global Publishing of Cape Town in 2007 and again as The Vortex Theory in 2015. As well as being a pioneering physicist and metaphysician, David is also a nutritionist, a stonemason, a ceremonialist and a nomadic troubadour. He is a father of nine and grandfather to fourteen.

Lightning Source UK Ltd.
Milton Keynes UK
UKHW011811260120
357634UK00001B/4

9 781928 234234